Building Science and Materials 2 Checkbook

M D W Pritchard

Butterworths
London Boston Sydney Wellington Durban Toronto

All rights reserved. No part of this publication may be reproduced or transmitted in any form or by any means, including photocopying and recording without the written permission of the copyright holder, application for which should be addressed to the publishers. Such written permission must also be obtained before any part of this publication is stored in a retrieval system of any nature.

This book is sold subject to the Standard Conditions of Sale of Net Books and may not be resold in the UK below the net price given by the Publishers in their current price list.

First published 1981

© Butterworth & Co (Publishers) Ltd 1981

British Library Cataloguing in Publication Data

Pritchard, Michael David William
  Building science and materials 2 checkbook.
  1. Building materials – Problems, exercises, etc
  I. Title
  691'.076   TA404.3   80-41176

ISBN 0-408-00640-4
ISBN 0-408-00607-2 Pbk

Typeset by Reproduction Drawings Ltd, Sutton, Surrey
Printed and Bound by Robert Hartnoll Ltd., Bodmin

Contents

Preface vii

1 Thermal studies 1
Thermal transmittance 1
Comparison of heating fuels 3
Calculation of internal temperature 4
Average U-values 5
Psychrometry 9
Exercises 15

2 Sound 18
Sound waves 18
Velocity of sound 19
Velocity, frequency and wavelength 21
Decibel scale 22
Sound insulation 27
Exercises 30

3 Applied mechanics 33
Stress 33
Strain 35
Modulus of electicity 35
Tensile strength 37
Beam reactions 39
Shear force 40
Bending moment 42
Internal forces in a truss 45
Internal forces in a beam 47
Internal forces in a column 50
Exercises 51

4 Water and building 54
Surface tension 54
Capillarity 56
Practical measures for preventing moisture penetration 58
Porosity of building materials 58
Electrolytic corrosion 60
Exercises 62

5 Cement, aggregates and concrete 65
Cements 65
Concrete aggregates 69
Concrete 75
Exercises 80

6 Plastics and paints   83
   Plastics   83
   Types of thermoplastics   86
   Types of thermosetting plastics   88
   Paints   88
   Emulsion paints   90
   Exercises   91

7 Timber, bricks, blocks and plaster   93
   Timber   93
   Insect attack   95
   Fungal attack   96
   Prevention and eradication of dry rot   97
   Wet rot   98
   Timber preservatives   98
   Bricks   99
   Concrete blocks   101
   Gypsum plasters   102
   Exercises   103

   Answers to exercises   106

   Index   109

Note to Reader

As textbooks become more expensive, authors are often asked to reduce the number of worked and unworked problems, examples and case studies. This may reduce costs, but it can be at the expense of practical work which gives point to the theory.

Checkbooks if anything lean the other way. They let problem-solving establish and exemplify the theory contained in technician syllabuses. The Checkbook-reader can gain *real* understanding through seeing problems solved and through solving problems himself.

Checkbooks do not supplant fuller textbooks, but rather supplement them with an alternative emphasis on illustration and an ample provision of worked and unworked problems. The brief outline of essential data—definitions, formulae, laws, regulations, codes of practice, standards, conventions, procedures, etc—will be a useful introduction to a course and a valuable aid to revision. Short-answer and multi-choice questions are a valuable feature of many Checkbooks, and answers are given whenever possible.

Checkbook authors are carefully selected. Most are experienced and successful technical writers; all are experts in their own subjects; but a more important qualification still is their ability to demonstrate and teach the solution of problems in their particular branch of technology, mathematics or science.

Authors, General Editors and Publishers are partners in this major low-priced series whose essence is captured by the Checkbook symbol of a question or problem 'checked' by a tick for correct solution.

Preface

The aim of this textbook is to provide information and assistance to students studying the TEC unit Science and Materials II on Building or Civil Engineering Courses.

It is hoped that the combination of worked examples, facts and problems will enable the reader to grasp the essential information and appreciate its application in the design and construction of buildings. Many of the examples have been chosen to show the application of the information and to assist the reader in future studies. Each chapter includes a set of exercises for the student to work out with answers given at the end of the book.

The author is indebted to Colin Bassett, the Series editor of the Building Checkbooks, for his help and encouragement throughout the preparation of this work.

Michael Pritchard
Guildford County College of Technology

Acknowledgement

Extracts from British Standards are reproduced by permission of British Standards Institution, 2 Park Street, London, W1A 2BS from whom complete copies can be obtained.

Butterworths Technical and Scientific Checkbooks

General Editor for Building, Civil Engineering, Surveying and Architectural titles:
Colin R. Bassett, lately of Guildford County College of Technology.

General Editors for Science, Engineering and Mathematics titles:
J. O. Bird and A. J. C. May, Highbury College of Technology, Portsmouth

A comprehensive range of Checkbooks will be available to cover the major syllabus areas of the TEC, SCOTEC and similar examining authorities. A full list is given below and classified according to levels.

Level 1 (Red covers)
Mathematics
Physical Science
Physics
Construction Drawing
Construction Technology
Microelectronic Systems
Engineering Drawing
Workshop Processes and Materials

Level 2 (Blue covers)
Mathematics
Chemistry
Physics
Building Science and Materials
Construction Technology
Electrical and Electronic Applications
Electrical and Electronic Principles
Electronics
Microelectronic Systems
Engineering Drawing
Engineering Science
Manufacturing Technology

Level 3 (Yellow covers)
Mathematics
Chemistry
Building Measurement
Construction Technology
Environmental Science
Electrical Principles
Electronics
Electrical Science
Mechanical Science
Engineering Mathematics and Science

Level 4 (Green covers)
Mathematics
Building Law
Building Services and Equipment
Construction Site Studies
Concrete Technology
Economics of the Construction Industry
Engineering Instrumentation and Control
Geotechnics

Level 5 (Purple covers)
Building Services and Equipment
Construction Technology

1 Thermal studies

Thermal transmittance

The most important concept in thermal studies, at this stage, is that of the **thermal transmittance**. The thermal transmittance is denoted by the symbol U and is often referred to as a U-value. It represents the heat flow rate in watts through one square metre of a construction when a temperature difference of 1° C exists between inside and outside temperatures. At a later stage it will be necessary to be more precise about the inside and outside temperatures since it is usual to use the environmental temperature for calculation purposes, this is a combination of the air temperature and the mean radiant temperature.

The U-value summarises as a single number the somewhat complex heat transfer mechanisms that occur through a construction. These are:
(i) convection and radiation exchanges at the inside and outside surfaces.
(ii) the conduction of heat through the solid parts of the construction.
(iii) the convection, conduction and radiation heat transfer across cavities in the construction.

The heat transfer at the outside surface will be dependent on the external conditions which will vary with the wind velocity and orientation. The heat conducted through the material of the construction will vary with the moisture content of the materials, being considerably higher when the moisture content is high. The heat transfer at the inside surfaces will depend on the orientation of the surface and the direction of heat flow.

It will be appreciated that the U-value will depend on the exposure of the building and the direction of heat flow through the construction. The CIBS Guide (Section A3) gives the U-values of many common forms of construction for different exposures and orientations.

Before proceeding to perform calculations it is essential to establish the units of the U-value. Referring to the definition above, it will be seen that the units of the U-value are W/m² °C.

Problem 1. An external cavity wall has a U-value of 0.96 W/m² °C. State the heat loss rate that occurs through 1 m² of this wall when a temperature difference of 1° C exists between the inside and outside temperatures.

It will be seen from the definition of the U-value that the heat loss rate is **0.96 W**.

Problem 2. Calculate the heat loss rate through 10 m² of the cavity wall specified in *Problem 1* when a temperature difference of 1° C exists between the inside and outside temperatures.

It is evident that as the area of the wall increases the heat loss rate increases proportionally. Thus heat loss rate = 10 × 0.96 = **9.6 W**.

> *Problem 3.* Calculate the heat loss through 10 m² of the cavity wall specified in *Problem 1* when a temperature difference of 18° C exists between the inside and outside temperatures.

Again it is clear that as the temperature difference increases the heat loss rate increases proportionally: Thus heat loss rate = 10 × 0.96 × 18 = **172.8 W**.

Heat loss rate

The above examples illustrate the method of calculating the heat loss rate through a construction. This can be summarised by the formula:

$$Q = AU(t_i - t_o)$$

Where Q is the heat loss rate in W; A is the area in m², U is the U-value in W/m² °C and $t_i - t_o$ is the temperature difference between the inside and outside temperatures.

In order to find the total heat loss rate for a building the heat loss rate through each element of the construction can be found and the total heat loss rate evaluated. This is illustrated in the next example.

> *Problem 4.* A detached bungalow 10 m by 12 m on plan has a ceiling height of 2.4 m. The external cavity walls have a U-value of 0.96 W/m² °C. The external walls include 18 m² of single glazing with a U-value of 5.6 W/m² °C. The solid floor construction has a U-value of 0.64 W/m² °C. The roof, which has a pitch of 30°, has a U-value of 1.5 W/m² °C.
> The average internal temperature is 19° C and the external temperature may be taken as -1° C.
>
> Calculate:
> (a) the heat loss rate.
> (b) the heat loss from the bungalow in a 24 hour period.
> (c) the cost of providing the heat for a 24 hour period by electrical heating when the cost of a unit of electricity is 6.3 p.

Before proceeding with the calculations the following points must be noted:
(i) the area of the cavity walls must be found making allowance for the area of the glazing. Thus the area of the cavity walls is given by:

$$2(10 \times 2.4 + 12 \times 2.4) - 18 = 87.6 \text{ m}^2$$

(ii) U-values for pitched roofs are conventionally quoted so that the appropriate area to be used is the plan area; in this case 10 m × 12 m giving 120 m².

The heat loss rate calculations for part (a) can now be tabulated as follows:

Element	Area (m²)	Heat loss rate $Q = AU(t_i - t_o)$
Cavity walls	87.6	$87.6 \times 0.96 \times (19 - (-1)) = 1681.9$
Windows	18	$18 \times 5.6 \times (19 - (-1)) = 2016$
Floor	120	$120 \times 0.64 \times (19 - (-1)) = 1536$
Roof	120	$120 \times 1.5 \times (19 - (-1)) = 3600$

Total heat loss rate = 8833.9 W

The total heat loss rate is found to be 8833.9 W, this result indicates a higher degree of precision than is warranted by the initial data and an approximated figure of 8850 W would seem reasonable. Note that 8850 W would usually be expressed as 8.85 kW.

In order to calculate the heat loss for a 24 hour period as required in part (b) of the question it is necessary to remember that 1 W is 1 J/s, thus multiplying the heat loss rate in watts by the time period in seconds will yield the heat loss in joules.

Heat loss = heat loss rate × time
= 8850 × 24 × 3600 = 764640000 J

It will be immediately obvious that the joule is a very small amount of heat and it is usual to express these answers either in megajoules or gigajoules as appropriate. Note that 1 megajoule is 10^6 joules and that 1 gigajoule is 10^9 joules. The appropriate abbreviations are MJ for megajoule and GJ for gigajoule. Thus the above heat loss can be written as 764.64 MJ.

It is unfortunate that energy is not yet sold by the megajoule. The unit of electricity charged is the kilowatt-hour; this being the product of the number of kilowatts and the number of hours of use.

For part (c) of the question the number of kilowatt-hours is 8.85 × 24 = 212.4. Hence the cost is 212.4 × 6.3 = 1338 p = £13.38.

Comparison of heating fuels

It is very instructive to compare the different fuels and the methods of pricing which are in common use. This is of course essential if cost comparisons are to be made.

Electricity The unit is the kilowatt-hour

1 kWh = 1 kW × 1 hour
= 1000 W × 3600 s = 3600000 J = 3.6 MJ.

Gas Gas is sold by the therm. 1 therm is 105.5 MJ. The practical situation is somewhat more complicated since most domestic gas meters are calibrated in hundreds of cubic feet.

The following method is currently in use for calculating the number of therms of gas used:

$$\text{Number of therms} = \frac{\text{cubic feet of gas} \times \text{calorific value}}{100\,000}$$

The calorific value being quoted in Btu/ft³ and having a typical value of 1035 for natural gas. Should the Gas Board decide to metricate their equipment this calorific value would be 38.6 MJ/m³.

Oil Oil is sold by the litre. Several grades of heating oil are available depending on the boiler plant employed. In order to determine the cost of the heat provided by the oil it is necessary to know both its calorific value, usually given in MJ/kg and its relative density. As an example: a heating oil has a relative density of 0.835 and a calorific value of 42.7 MJ/kg. From the relative density value the mass of 1 litre of the oil is 0.835 kg and thus the heat produced by 1 litre of this oil is 0.835 × 42.7 = 35.65 MJ.

Solid fuel Coal is sold by the tonne: the calorific value depends on the grade of the coal and a value about 27 MJ/kg would be appropriate for general purpose coal.

> *Problem 5.* For the bungalow detailed in *Problem 4* determine for a 24-hour period:
> (a) the number of therms of gas used assuming a boiler efficiency of 75%.
> (b) the number of litres of oil burnt by a boiler using an oil with a relative density of 0.79 and a calorific value of 43.4 MJ/kg. The boiler efficiency is 75%.
> (c) the number of kilograms of coal of calorific value 27 MJ/kg assuming a boiler efficiency of 60%.

In *Problem 4* it was found that the heat loss in the 24-hour period was 764.64 MJ.

In order to determine the number of therms of gas used it must be remembered that 1 therm = 105.5 MJ.

Number of therms of gas = 764.64/105.5 = 7.25 therms

This would give the number of therms used if the boiler was 100% efficient, as this is not the case more gas must be used in order to provide the heat required.

Number of therms of gas = 7.25 × 100/75 = **9.67 therms**

In calculating the number of litres of oil burnt the heat produced by 1 litre of oil should be calculated as follows:

Heat produced by 1 litre of oil = 0.79 × 43.4 = **34.3 MJ**

If the boiler were 100% efficient the number of litres of oil used would be given by:

Number of litres = 764.64/34.3 = 22.3 litres

As the boiler has an efficiency of 75% the number of litres of oil actually used is:

Number of litres = 22.3 × 100/75 = **29.7 litres**

For the case of the coal fired boiler it is readily seen that:

$$\text{Quantity of coal} = \frac{764.64}{27} \times \frac{100}{60} = \textbf{47.4 kg}$$

Calculation of internal temperature

In studies of the occurrence of mould growth and condensation it is often necessary to calculate the internal temperature in a dwelling when the heat input is specified. The following example illustrates a typical calculation.

Problem 6. The table below gives the properties of a centre-terraced house:

Component	Area (m^2)	U-value (W/m^2 °C)
Cavity wall	32.4	0.96
Window	21.5	5.6
Roof	46.75	1.5
Floor	46.75	0.64

The heat losses through the party walls may be ignored in this example.

It is required to calculate the average internal temperature when the outside temperature is $-1°$ C and the heat input to the house is 4 kW.

Denote the internal temperature by t_i and then the following expressions for the heat loss rates may be written down:

Heat loss rate through cavity wall $= 32.4 \times 0.96 \times (t_i - (-1))$
$= 32.4 \times 0.96 \times (t_i + 1)$
$= 31.10 (t_i + 1)$ watts

Heat loss rate through windows $= 21.5 \times 5.6 \times (t_i - (-1))$
$= 120.4 (t_i + 1)$ watts

Heat loss rate through roof $= 46.75 \times 1.5 \times (t_i - (-1))$
$= 70.13 (t_i + 1)$ watts

Heat loss rate through floor $= 46.75 \times 0.64 (t_i - (-1))$
$= 29.92 (t_i + 1)$ watts

The total heat loss rate must balance the heat input to the house of 4 kW or 4000 W. Hence:

$31.10 (t_i + 1) + 120.4 (t_i + 1) + 70.13 (t_i + 1) + 29.92 (t_i + 1) = 4000$
$251.55 (t_i + 1) = 4000$
$t_1 + 1 = 4000/251.55$
$t_i + 1 = 15.90$
$t_i \quad = 14.9°$ C

Thus the internal temperature is found to be $14.9°$ C on average throughout the house.

Average U-values

The Building Regulations (1978) specify standards of insulation for new buildings. For domestic dwellings maximum U-values for floors, walls and roofs are specified. A maximum value for the average U-value for perimeter walling for domestic dwellings is also specified. Assume that the perimeter wall is comprised of a number of different constructions having different areas and U-values as follows:

Construction 1 of area A_1 and U-value U_1
Construction 2 of area A_2 and U-value U_2

Then the average U-value = $\dfrac{A_1 U_1 + A_2 U_2}{A_1 + A_2}$

The formula is readily extended to any number of different constructions. The following examples illustrate the type of calculations that are involved in interpreting the Building Regulations.

> *Problem 7.* A detached bungalow is 10 m by 14 m on plan and has a ceiling height of 2.5 m. The perimeter walls are of a cavity construction with a U-value of 0.96 W/m² °C. The perimeter walls include 25 m² of single glazed windows having a U-value of 5.7 W/m² °C. Determine the average U-value of the perimeter wall.

The area of the cavity wall construction is readily found making a deduction for the area of the windows.

Area of cavity wall = 2 (10 × 2.5 + 14 × 2.5) − 25 = 95 m²

Thus using the formula given above:

$$\text{Average } U\text{-value} = \frac{95 \times 0.96 + 25 \times 5.7}{95 + 25}$$

$$= \frac{233.7}{120} = 1.95 \text{ W/m}^2 \text{ °C}$$

> *Problem 8.* A semi-detached bungalow is 12 m by 10 m on plan and has a ceiling height of 2.5 m. The bungalow is attached to its neighbour on the 12 m side. The external cavity wall has a U-value of 0.9 W/m² °C. The external wall includes 20 m² of single glazed windows having a U-value of 5.7 W/m² °C. The U-value of the party wall is to be taken as 0.5 W/m² °C. Determine the average U-value of the perimeter walls.

The areas involved are as follows:

Area of cavity wall = 2.5 × (10 + 10 + 12) − 20 = 60 m²
Area of party wall = 2.5 × 12 = 30 m²
Area of glazing = 20 m²

The average U-value is found by extending the formula given above to three materials.

$$\text{Average } U\text{-value} = \frac{60 \times 0.9 + 30 \times 0.5 + 20 \times 5.7}{60 + 30 + 20}$$

$$= \frac{183}{110} = 1.66 \text{ W/m}^2 \text{ °C.}$$

Note that the correct method of writing the formula for the average U-value where n types of construction are involved is:

$$\text{Average } U\text{-value} = \frac{\sum_{i=1}^{n} A_i U_i}{\sum_{i=1}^{n} A_i}$$

Problem 9. A semi-detached house is 10 m × 7 m on plan and is attached on the 10 m side. The external walls have a height of 5.2 m. The U-values of the cavity brickwork and the party wall are 0.96 W/m² °C and 0.5 W/m² °C respectively. Determine the permissible area of single glazing having a U-value of 5.7 W/m² °C if the average U-value of the perimeter walls is not to exceed 1.8 W/m² °C.

It is necessary in the first instance to write down the relevant areas:

Let the area of the single glazing = A m²
Then the area of the cavity wall = 5.2 × (10 + 7 + 7) − A
= 124.8 − A
Area of party wall = 10 × 5.2
= 52 m²
Total area of perimeter wall = 124.8 + 52
= 176.8 m²

Using the formula for the average U-value:

$$\text{Average } U\text{-value} = \frac{(124.8 - A) \times 0.96 + 52 \times 0.5 + A \times 5.7}{176.8}$$

$$= \frac{119.8 - 0.96A + 26 + 5.7A}{176.8}$$

$$= \frac{145.8 + 4.74A}{176.8}$$

It is required that the average U-value must not exceed 1.8 W/m² °C. Hence:

$$\frac{145.8 + 4.74A}{176.8} \leqslant 1.8$$

$$145.8 + 4.74A \leqslant 1.8 \times 176.8$$
$$145.8 + 4.74A \leqslant 318.24$$
$$4.74A \leqslant 318.24 - 145.8$$
$$4.74A \leqslant 172.44$$
$$A \leqslant \frac{172.44}{4.74}$$
$$A \leqslant 36.38 \text{ m}^2$$

Thus the area of single glazing must be less than or equal to 36.38 m².

Problem 10. The perimeter walls of a detached dwelling consist of 100 m² of cavity wall and 18.75 m² of single glazing. The U-value of the glazing may be taken as 5.7 W/m² °C. Determine the necessary U-value of the cavity wall if the average U-value is not to exceed 1.8 W/m² °C and the maximum U-value of the cavity wall is to be 1.0 W/m² °C.

Let U_c be the U-value of the cavity wall. Then using the usual formula for the average U-value and the restriction that this must not exceed 1.8 W/m² °C the following expression can be written down:

$$\frac{100U_c + 18.75 \times 5.7}{100 + 18.75} \leqslant 1.8$$

$$\frac{100U_c + 106.88}{118.75} \leqslant 1.8$$

$$100U_c + 106.88 \leqslant 1.8 \times 118.75$$
$$100U_c + 106.88 \leqslant 213.75$$
$$100U_c \leqslant 213.75 - 106.88$$
$$100U_c \leqslant 106.87$$
$$U_c \leqslant 1.0687 \text{ W/m}^2 \text{ °C}$$

This value of U_c would ensure that the average U-value did not exceed 1.8 W/m² °C; however the maximum U-value of the cavity wall is 1.0 W/m² °C. Thus the second restriction is the critical one and the maximum U-value of the cavity wall is 1.0 W/m² °C.

> *Problem 11.* A semi-detached house has external cavity walls with a U-value of 0.8 W/m² °C. The party wall, which constitutes 30% of the area of the perimeter walls, has a U-value of 0.5 W/m² °C. The wall between the house and the attached garage, which constitutes 20% of the area of the perimeter walls, has a U-value of 1.7 W/m² °C. Determine the percentage of the perimeter walls that can be single glazing with a U-value of 5.7 W/m² °C if the average U-value of the perimeter walls is not to exceed 1.8 W/m² °C.

The external cavity wall and the windows together form 50% of the perimeter wall.

Let the percentage of single glazing be x per cent, then the percentage of external cavity walling is $(50 - x)$. Using the percentage areas the average U-value can be expressed as:

$$\frac{30 \times 0.5 + 20 \times 1.7 + 0.8(50 - x) + 5.7x}{100}$$

Simplifying this:

$$\text{Average } U\text{-value} = \frac{15 + 34 + 40 - 0.8x + 5.7x}{100}$$

$$= \frac{89 + 4.9x}{100}$$

If this is not to exceed 1.8 W/m² °C then

$$\frac{89 + 4.9x}{100} \leqslant 1.8$$
$$89 + 4.9x \leqslant 180$$
$$4.9x \leqslant 91$$
$$x \leqslant 18.6$$

Thus the area of single glazing must be less than 18.6 per cent of the perimeter walls.

If this is expressed as a percentage of the external wall excluding the party and garage wall: the percentage of single glazing is given as

$$\frac{18.6}{100} \times \frac{100}{(100 - 50)}$$

since the party wall and garage wall together constitute 50% of the perimeter walls.

Thus percentage of glazing expressed as a percentage of the external wall area is 37%.

Psychrometry

Psychrometry, often referred to as hygrometry, is the study of mixtures of air and water vapour. This subject is of considerable importance in the design of air conditioning systems and the prediction of condensation in buildings. Before proceeding to calculations it is essential that the basic physical processes are understood and these will be considered in outline.

EVAPORATION OF WATER

Consider a container, as illustrated in *Fig 1(a)*, of fixed volume and maintained at a constant temperature $T°$ C. Suppose, initially, that all the air is removed and the pressure is thus zero. If a small quantity of water is introduced this will evaporate to occupy the complete volume of the container and the molecules of water vapour will create a vapour pressure, e, as shown in *Fig 1(b)*.

As more water is introduced, the evaporation of this water causes the vapour pressure

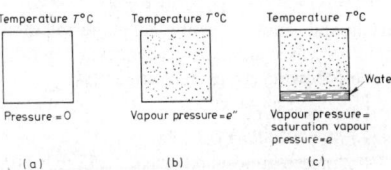

Fig 1 Evaporation of water

to increase until a maximum vapour pressure is reached. This vapour is then said to be saturated and the vapour pressure is called the **Saturation Vapour Pressure** (s.v.p.), denoted by e''. If more water is introduced then it will not evaporate but will exist as water, as illustrated in *Fig 1(c)*.

If the temperature of the container is raised then more water will evaporate and the pressure in the container will increase. Thus the s.v.p. increases with temperature. Values of the s.v.p. are given in *Table 1* for a range of temperatures.

In the above discussion the container was initially evacuated, but the water would evaporate in exactly the same way if the container was originally filled with air. **Dalton's Law of Partial Pressures** states that the mixture of water vapour and air behaves if each constituent occupied the whole container at the same temperature. More specifically Dalton's Law states that the total pressure in the container is the sum of the air pressure and the water vapour pressure. Hence the mixture can be regarded as composed of dry air and water vapour acting independently.

Values obtained for the s.v.p. can be used to predict the behaviour of mixtures of air and water vapour.

The foregoing assumes that Dalton's Law applies accurately to vapours, although this is not strictly true, the divergence under normal atmospheric conditions is negligible.

TABLE 1 Saturation vapour pressure over water

Temperature (°C)	Saturation vapour pressure (kPa)
0	0.61078
2	0.70547
4	0.81294
6	0.93465
8	1.0722
10	1.2272
12	1.4017
14	1.5977
16	1.8173
18	2.0630
20	2.3373
22	2.6430
24	2.9831
26	3.3608
28	3.7796
30	4.2430

UNITS OF VAPOUR PRESSURE

The air and vapour pressures are expressed in kilopascals (kPa). Other units that are commonly used are newtons per square metre (N/m^2) and the millibar (mb).

The following conversions are useful:

1 Pa = 1 N/m^2
1 mb = 100 Pa = 0.1 kPa
1 standard atmosphere = 1013.25 mb = 101.325 kPa

METHODS OF SPECIFYING THE HUMIDITY OF THE AIR

In situations existing in construction, the water vapour in the air is not normally saturated. A profusion of methods exist for specifying the humidity. The following definitions and formulae are in accordance with BS 1339:1965 and these may well be found not to agree with other publications.

Absolute humidity, d_v is the mass of water vapour present in unit volume of moist air and is expressed in kg/m^3.

Specific humidity, q is the mass of water vapour present in unit mass of moist air and is expressed in kg/kg moist air.

Mixing ratio, r is the ratio of the mass of water vapour to the mass of dry air with which the water vapour is associated and is usually expressed as kg/kg dry air.

The above definitions compare the mass of water vapour in the air with either the **volume** of air containing it or the **mass** of air containing it. Similar definitions are found for specifying the moisture contents of building materials, the usual definition of moisture content corresponding to the definition of the mixing ratio.

In many cases it is more useful to compare the water vapour present in the mixture to the amount of water vapour present when saturated.

Relative Humidity, U_w is defined as:

$$\frac{\text{actual vapour pressure}}{\text{s.v.p. at the same temperature}} \times 100\% = \frac{e}{e''} \times 100\%$$

Percentage Saturation is defined as:

$$\frac{\text{mixing ratio}}{\text{mixing ratio at saturation at the same temperature and pressure}} \times 100\%$$

If the mixture behaved as a perfect gas the percentage saturation and the relative humidity would be identical. Under ordinary climatic conditions the difference is small and seldom exceeds 2%.

As the water vapour pressure and the mass of water vapour present in the air are clearly related the above definitions are also related. The following formulae are adequate for most practical circumstances.

Mixing ratio r

$$r = \frac{0.622e}{P - e}$$

where e is the vapour pressure and P the total atmospheric pressure.

Absolute humidity, d_v

$$d_v = \frac{2.17e}{T}$$

where e is the vapour pressure in kPa.
T is the absolute temperature in °K
and d_v is in kg/m³ of moist air.

Percentage Saturation is related to the relative humidity by the equation:

$$\frac{\text{percentage saturation}}{\text{relative humidity}} = \frac{P - e''}{P - e}$$

where e is the vapour pressure and e'' is the s.v.p. at the same temperature.

Saturation vapour pressure. *Table 1* gives accurate values of the s.v.p. but in some circumstances it is advantageous to be able to calculate the s.v.p. at any given temperature. A close approximation to the s.v.p. can be obtained from the following formula.

$$\log_{10} e'' = \frac{7.5t}{237.3 + t} - 0.21429$$

where e'' is the s.v.p. in kPa and t is the temperature in °C.

Problem 12. A mixture of air and water vapour at 16° C under an atmospheric pressure of 101.3 kPa has a water vapour pressure of 1.2 kPa. Determine
(a) the relative humidity;
(b) the mixing ratio;
(c) the absolute humidity;
(d) percentage saturation.

The vapour pressure $e = 1.2$ kPa.
From *Table 1* the s.v.p. at $16°$ C is seen to be: $e'' = 1.8173$ kPa.

The relative humidity $= \dfrac{e}{e''} \times 100\%$

$$= \dfrac{1.2}{1.8173} \times 100\% = \mathbf{66.03\%}$$

The mixing ratio is found from the formula:

$r = \dfrac{0.622e}{P - e}$

$= \dfrac{0.622 \times 1.2}{101.3 - 1.2} = \mathbf{0.007457 \text{ kg/kg dry air}}$

The absolute humidity is found from the formula:

$$d_v = \dfrac{2.17e}{T}$$

Note that T is in degrees Kelvin which are obtained by adding 273.15 to the temperature in degrees Celsius

$T = 273.15 + 16 = 289.15°$ K

$d_v = \dfrac{2.17 \times 1.2}{289.15} = \mathbf{0.009 \text{ kg/m}^3}$

The percentage saturation is calculated using

$\dfrac{\text{percentage saturation}}{\text{relative humidity}} = \dfrac{P - e''}{P - e}$

$\dfrac{\text{percentage saturation}}{66.03} = \dfrac{101.3 - 1.8173}{101.3 - 1.2}$

$\dfrac{\text{percentage saturation}}{66.03} = 0.9938$

percentage saturation $= 0.9938 \times 66.03 = \mathbf{65.62\%}$

Problem 13. Calculate the s.v.p. at a temperature of $16°$ C. Compare this with the value given in *Table 1*.

The s.v.p. is calculated using the formula:

$\log_{10} e'' = \dfrac{7.5t}{237.3 + t} - 0.21429$

When $t = 16°$ C

$\log_{10} e'' = \dfrac{7.5 \times 16}{237.3 + 16} - 0.21429$

$\log_{10} e'' = 0.259457$

In order to find e'' it is necessary to take the antilogarithm of the above value giving:

$e'' = 1.8174$ kPa

It will be observed that this is in close agreement with the value given in *Table 1*.

DEW POINT TEMPERATURE, t_d

This is the temperature at which the water vapour pressure is equal to the s.v.p. If the temperature is reduced below this value condensation of water will occur.

> *Problem 14.* Find the dew point temperature for water vapour at a pressure of 1.2 kPa.

An inspection of *Table 1* shows that a s.v.p. of 1.2 kPa will occur when the temperature is between 8° C and 10° C.
A more accurate estimate can be obtained by linear interpolation:

$$t_d = 8 + \frac{1.2 - 1.0722}{1.2272 - 1.0722} \times (10 - 8)$$

$t_d = 9.65\,°C$

The following approach is very instructive and the reader is advised to perform the method very carefully. Firstly it is necessary to construct a graph of the s.v.p. and temperature values given in *Table 1*. This graph is illustrated in *Fig 2*. The dew point temperature can then be read directly corresponding to a s.v.p. of 1.2 kPa. The dew point temperature is found to be 9.7° C.

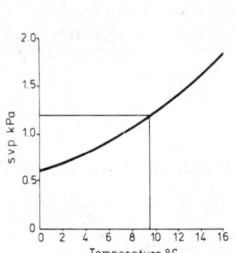

Fig 2 Dew point temperature (Probelm 14)

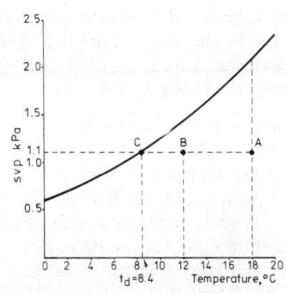

Fig 3 Determination of relative humidity (Problem 15)

> *Problem 15.* Moist air at 18° C has a water vapour pressure of 1.1 kPa. Determine the relative humidity. If this moist air is cooled to 12° C, determine the new value of the relative humidity. Find the dew point temperature of this moist air.

The easiest way to approach this problem is to use the graph of s.v.p. against temperature constructed from the values given in *Table 1*.

The initial conditions of the air, which has a temperature of 18° C and a water vapour pressure of 1.1 kPa are shown by the point A on *Fig 3*. The s.v.p. at 18° C is 2.063 kPa.

Thus the relative humidity $U_w = 100 \times \dfrac{1.1}{2.063} = 53.3\%$

If the moist air is cooled to 12° C there is no change in the amount of moisture present or in the water vapour pressure so the state of the moist air is shown by the point B on *Fig 3*. However at 12 °C the s.v.p. is reduced to 1.4017 kPa.

Hence relative humidity $U_w = 100 \times \dfrac{1.1}{1.4017} = 78.5\%$

It will be noted that the relative humidity has increased.

If the moist air is further cooled the point C on *Fig 3* will be reached. At this s point the vapour pressure is equal to the s.v.p. and the temperature is the dew point temperature which can be read as 8.4° C.

Since the water vapour pressure and the s.v.p. are equal the relative humidity is 100%.

MEASUREMENT OF RELATIVE HUMIDITY

The wet and dry bulb hygrometer is a very common instrument for measuring relative humidity. As water evaporates from the wick around the wet bulb the extraction of latent heat reduces the temperature of the wet bulb below that of the dry bulb. The rate of evaporation depends on the existing water vapour pressure in the surrounding air being greater for low vapour pressures. The wet bulb temperature depends on the rate of evaporation and thus the wet bulb and dry bulb readings enable the relative humidity to be found. This is usually done using suitable tables but the following method allows the relative humidity to be calculated with reasonable accuracy. The actual vapour pressure is calculated from the wet bulb and dry bulb temperatures using one of the following formulae:

$e = e' - 6.66 \times 10^{-4} P(t - t')$
or $e = e' - 7.99 \times 10^{-4} P(t - t')$

where e = vapour pressure in kPa;

e' = s.v.p. at the wet bulb temperature. This can be calculated using the previously given formula for s.v.p. or from *Table 1*;

P = atmospheric pressure in kPa;

t = dry bulb temperature in °C;

t' = wet bulb temperature in °C.

The first formula is used for forced-ventilated hygrometers of the sling or Assmann type and is applicable from 0° C to 150° C. The second formula is used for hygrometers in a screen and is applicable from 0° C to 60° C.

Problem 16. An Assmann hygrometer gave wet and dry bulb readings of 14 °C and 18 °C respectively. Assuming the atmospheric pressure to be 101 kPa determine the relative humidity of the air.

From *Table 1*:
s.v.p. at wet bulb temperature $e' = 1.5977$ kPa
s.v.p. at dry bulb temperature $e'' = 2.0630$ kPa

Since the hygrometer is forced-ventilated the actual vapour pressure is calculated using

$e = e' - 6.66 \times 10^{-4} P(t - t')$
$e = 1.5977 - 6.66 \times 10^{-4} \times 101 (18 - 14)$
$\quad = 1.3286 \text{ kPa}$

The relative humidity $U_w = 100 \dfrac{e}{e''}$

Note that e'' is the s.v.p. at the dry bulb temperature.

$$U_w = 100 \times \frac{1.3286}{2.0630} = 64.4\%$$

Exercises (*Answers on page 106*)

1 Calculate the heat loss rate through an external wall having dimensions 2.3 m by 6 m and a U-value of 0.84 W/m² °C when a temperature difference of 15° C exists between the inside and outside temperatures.

2 An external wall has dimensions 2.5 m by 5 m. Find the U-value if the heat loss rate is 200 W when the inside temperature is 17° C and the outside temperature is −1° C.

3 A window has a U-value of 5.6 W/m² °C and an area of 6.5 m². Determine the outside temperature which would cause a heat loss rate of 600 W when the inside temperature is 15° C.

4 A roof, 3 m by 4 m on plan is pitched at an angle of 30°. The U-value of the roof is 1.4 W/m² °C. Determine the heat loss rate when the difference between the inside and outside temperatures is 14.5° C.

5 A detached house is 12 m by 9 m on plan. The height of the external walls is 5 m, and they contain 20% single glazing. Calculate the heat loss rate from the house if the inside and outside temperatures may be taken as 16° C and −2° C respectively. The following U-values may be assumed:
external wall   0.80 W/m² °C   roof   1.5 W/m² °C
single glazing   5.6 W/m² °C   floor   0.39 W/m² °C

6 The heat loss from a building is 7 kW on average over a 24 hour period.
  (a) If the heating is provided electrically, find the cost of the electricity. Assume that 1 unit of electricity costs 6.3 pence.
  (b) If the heating is provided by an oil fired boiler find the cost of the oil used, assuming the following data:
  cost of oil   25 pence per litre   calorific value   43.5 MJ/kg
  relative density of oil   0.79   boiler efficiency   70%

7 A semidetached bungalow is 10 m by 11 m on plan and is attached on the 10 m side. The storey height of the bungalow is 2.5 m. The external walls contain 20 per cent double glazing. The gas bill for heating for a November was £14.74. The following data may be assumed:
U-values:
external wall   0.8 W/m² °C   roof   0.6 W/m² °C
double glazing   2.8 W/m² °C   cost of gas   33 pence per therm
party wall   0.5 W/m² °C   efficiency of gas boiler 85%
floor   0.36 W/m² °C

Calculate:
(a) number of therms used;
(b) average rate of heat use;
(c) the average temperature difference throughout November between the inside and outside temperatures.

8 A semi-detached house has the following properties

Component	Area (m^2)	U-value (W/m^2 °C)
External walls	80	0.96
Windows	20	5.60
Party wall	45	0.50
Floor	49.5	0.76
Roof	49.5	0.60

Calculate the average internal temperature when the heat input is 3 kW and the outside temperature is 1° C.

9 Find the average U-value for a construction which consists of: 20 m^2 of cavity wall having a U-value of 0.8 W/m^2 °C and 8 m^2 of double glazing having a U-value of 2.8 W/m^2 °C.

10 A centre terraced house is 9 m by 6.5 m on plan and is attached on the 9 m sides. The height of the first floor ceiling above the ground floor is 4.9 m. The external walls are of a cavity construction (U = 0.90 W/m^2 °C) and contain 40% single glazing (U = 5.7 W/m^2 °C). The U-value of the party walls may be taken as 0.5 W/m^2 °C. Determine the average U-value of the perimeter walls.

11 The external walls of a detached building are of cavity construction having a U-value of 0.8 W/m^2 °C. Determine the maximum percentage of single glazing (U = 5.7 W/m^2 °C) that is permissible if the average U-value is not to exceed 1.8 W/m^2 °C.

12 Repeat Exercise 11 for double glazing (U = 2.8 W/m^2 °C).

13 Repeat Exercise 11 for U-value of the cavity construction ranging from 0.5 to 1.0 W/m^2 °C. in steps of 0.1. Construct a suitable graph to show the variation of percentage single glazing against the U-value of the cavity wall.

14 The perimeter walls of a building are:
Party wall: 25% of area of perimeter wall. U-value = 0.5 W/m^2 °C
Wall between house and partly ventilated space: 15% of area of perimeter wall. U-value = 1.5 W/m^2 °C
External cavity wall: U-value 0.85 W/m^2 °C
Single glazing: U-value 5.7 W/m^2 °C
Determine the maximum percentage single glazing in the external cavity wall that is permissible if the average U-value of the perimeter walls is not to exceed 1.8 W/m^2 °C.

15 Repeat Exercise 14 for a double glazing having a U-value of 2.8 W/m^2 °C.

16 The U-value for a glazing system is given, approximately, by

$$U = \frac{1}{0.179 + (n-1) \times 0.115}$$

where $n = 1$ for single glazing; $n = 2$ for double glazing, etc. Find the U-values for single, double, triple and quadruple glazing systems.

If these glazing systems are to be installed in external walls having a U-value of 0.85 W/m² °C determine, in each case, the maximum permissible percentage of glazing if the U-value is not to exceed 1.8 W/m² °C.

17 A mixture of air and water vapour is at a temperature of 18° C and a total pressure of 100 kPa. The water vapour pressure is 1.3 kPa. Determine:
    (i) the relative humidity;
    (ii) the mixing ratio.

18 Air at 18 °C and 50 % relative humidity is cooled to 14 °C. Determine the new relative humidity.

19 A mixture of air and water vapour at 16° C has a mixing ratio of 0.009 kg/kg dry air. Assuming the atmospheric pressure to be 101.3 kPa calculate:
    (i) the vapour pressure;
    (ii) the relative humidity;
    (iii) the percentage saturation.

20 The air outside a building is at 0° C and 100% relative humidity. Inside the building the air temperature is 20° C. The activities of the occupants of the building cause the mixing ratio of the inside air to exceed that of the outside air by 0.0034 kg/kg dry air. Determine:
    (i) the mixing ratio of the outside air;
    (ii) the mixing ratio of the inside air;
    (iii) the relative humidity of the inside air.
    Assume the atmospheric pressure is 101.3 kPa.

21 Calculate the saturation vapour pressure at a temperature of 17° C. Compare this with the value obtained by interpolating *Table 1*.

22 The air in a room is at a temperature of 18° C and has a relative humidity of 60%. Determine the surface temperature of a window which just causes condensation to occur.

23 It was noted that condensation began to occur on the inside of a window when the surface temperature of the window was 7° C. The air temperature in the room was 18° C. Determine the relative humidity of the air in the room.

24 The wet and dry bulb temperatures recorded by an Assmann hygrometer were 20° C and 16° C respectively. The atmospheric pressure was 101.3 kPa. Determine the relative humidity and dew point temperature of the air.

2 Sound

Sound waves

Sound waves are a particular form of elastic waves. Elastic waves can occur in a medium having both mass and elasticity. If the medium has mass then a displaced particle can transfer momentum, and hence energy to an adjacent particle. The elasticity of the medium tends to return the displaced particle to its original position. It can thus be inferred that a medium is necessary for the propagation of sound waves.

In many cases the medium through which the sound wave is propagated is air. That air has mass is readily appreciated since it has a density of approximately 1.25 kg/m^3. A simple demonstration of the elasticity of air is furnished by a bicycle tyre pump. If a finger is placed over the outlet of the pump and the handle is pushed in and then released the handle returns back towards it original position.

These ideas are clearly demonstrated by considering a piston moving backwards and forwards in a tube full of air. In *Fig 1(a)* the piston is at rest and the spacing of the air molecules is uniform along the tube. If the piston moves forward, as in *Fig 1(b)*, the air molecules adjacent to the piston are compressed, and energy is stored in the compressed air. Due to the elasticity of the air the molecules endeavour to regain their original spacing. They cannot move back, since the piston is still there, so they move to the right along the tube. In doing so they acquire momentum and strike further molecules transferring their momentum to other molecules along the tube and causing them to become compressed. In this way the compression passes to the right along the tube. Suppose, now, that the piston returns to its original position, the air molecules near the piston assume their original spacing. This situation is shown in *Fig 1(c)*.

If the piston now moves to the left of its original position the separation between the molecules is increased, and a rarefaction is created. The molecules to right of this rarefaction are thus moved to the left as the molecules endeavour to regain their original spacing. In this manner the rarefaction is propagated to the right.

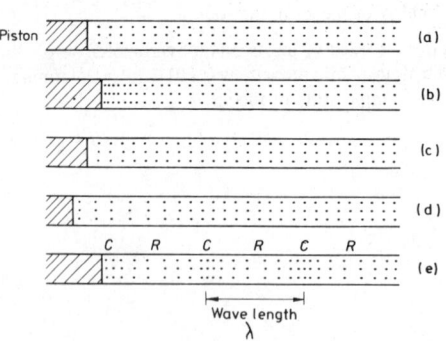

Fig 1 Propagation of a sound wave

If the piston moves backwards and forwards about its original position then an alternate system of compressions and rarefactions will be propagated down the tube as shown in *Fig 1(e)*. The letters C and R denoting compressions and rarefactions respectively.

It will be observed that the air molecules do not travel down the tube but oscillate about their original position.

Velocity of sound in air

The compressibility of air is related to its pressure and the mass of the air is given by its density. The following formula is applicable for calculating the velocity of sound in air:

$$v = \sqrt{\frac{1.4P}{\rho}}$$

where v = velocity of sound in air in m/s;
P = the air pressure in Pa;
ρ = density of air in kg/m^3.

Problem 1. Calculate the velocity of sound in air at 0° C when the atmospheric pressure is 101.3 kPa and the density of air is 1.293 kg/m^3.

$$v = \sqrt{\frac{1.4 \times 101.3 \times 10^3}{1.293}} = 331.2 \text{ m/s}$$

For air the ratio of P/ρ is nearly independent of pressure since if the pressure is doubled the density is doubled so that the velocity of sound in air changes very little with change in barometric pressure. However temperature does significantly effect the density of the air so that the velocity of sound in air varies with temperature. The velocity of sound in air for normal temperatures is given, approximately, by the relationship:

$v = 331.3 + 0.6t$ where t is the air temperature in °C.

Problem 2. Find the velocity of sound in air at 20° C.

$v = 331.3 + 0.6 \times 20 = 343.3$ m/s

Velocity of sound in liquids and solids

The velocity of sound in liquids and solids is much greater than the velocity in air. Some typical values are given below:

water at 20° C	1484 m/s
concrete	4250 to 5250 m/s
steel	5900 to 6100 m/s

APPLICATIONS OF VELOCITY OF SOUND

A knowledge of the velocity of sound has many applications in architectural acoustics and in non destructive testing of materials. The following examples will illustrate some of these applications.

Problem 3. A person at a sporting event is 10 m from one loudspeaker of the public address system and 25 m from another loudspeaker. If both loudspeakers produce the same sound at the same instant of time calculate:
(i) the time taken for the sound to reach the person from the nearest loudspeaker;
(ii) the time delay before the sound reaches the person from the second loudspeaker.
Assume the velocity of sound in air to be 340 m/s.

It will be remembered that:

$$\text{Time} = \frac{\text{Distance}}{\text{Velocity}}$$

Thus the time for the sound to reach the person from the nearest loudspeaker is:

$$\text{Time} = \frac{10}{340} = 0.0294 \text{ s}$$

The sound from the second loudspeaker has to travel a further 15 m, so the time delay is:

$$\text{Time} = \frac{15}{340} = 0.0441 \text{ s}$$

Problem 4. In a site test to measure the depth of a crack in a beam the time required for an ultrasonic pulse to be transmitted is measured. In the first instance the transducers are placed in positions A–A as shown in *Fig 2*. The time for the pulse to be transmitted through the 300 mm of concrete was found to be 66.67 μs.

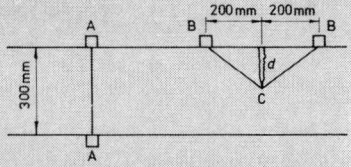

Fig 2 Measurement of crack depth (Problem 4)

The transducers were then placed in positions B–B and the time for the pulse to be transmitted was found to be 95.7 μs. Determine
(i) the velocity of sound in the concrete;
(ii) the depth of the crack.

With the transducers in positions A–A the pulse is transmitted directly and the velocity is given by

$$\text{Velocity} = \frac{\text{distance}}{\text{time}}$$

$$= \frac{0.300}{66.67 \times 10^{-6}} = 4500 \text{ m/s}$$

Note that 1 μs is 1×10^{-6} s.

In the second position the pulse will not cross the crack but will follow the path BCB as shown in *Fig 2*.

Let the depth of the crack be d metres. Then by using Pythagoras' theorem

$$BC = \sqrt{(d^2 + 0.2^2)}$$

The total path length is twice this, thus:

total path length = $2\sqrt{(d^2 + 0.2^2)}$

The distance is given by:

distance = velocity × time

The velocity is known, from above, to be 4500 m/s. Hence:

$$2\sqrt{(d^2 + 0.2^2)} = 4500 \times 95.7 \times 10^{-6}$$

$$\sqrt{(d^2 + 0.2^2)} = \frac{4500 \times 95.7 \times 10^{-6}}{2}$$

$$= 0.2153$$

On squaring both sides:

$d^2 + 0.2^2$	$= 0.2153^2$
$d^2 + 0.04$	$= 0.04636$
d^2	$= 0.00636$
d	$= 0.08$ m
d	$= 80$ mm

Thus the depth of the crack is **80 mm**.

Velocity, frequency and wavelength

Referring again to *Fig 1* the number of times in one second the piston completes a full cycle of movement is called the frequency which is denoted by f and has the unit Hertz (Hz). An example of a complete cycle of movement is from the extreme right hand position as shown in *Fig 1(b)* through the positions shown in *Figs 1(c)* and *1(d)* and back to the position in *Fig 1(b)*.

The wavelength of the sound is the distance between two successive compressions, or rarefactions, as illustrated in *Fig 1(e)* and is denoted by λ. The relationship between the velocity, frequency and wavelength can be deduced as follows.

The piston in *Fig 1* makes f complete cycles of movement in 1 s. Thus the time for one complete cycle is $1/f$ seconds. In this time a compression has travelled a distance of one wavelength. Hence the velocity can be found:

$$\text{velocity} = \frac{\text{distance}}{\text{time}} = \frac{\lambda}{1/f} \text{ which simplifies to give } v = f\lambda$$

Problem 5. Find the wavelength of a sound having a frequency of 1000 Hz if the velocity of sound in air is 340 m/s.

Rearranging the above formula gives:

$$\lambda = \frac{v}{f} = \frac{340}{1000} = 0.34 \text{ m}$$

Problem 6. Find the frequency of an ultrasonic wave whose wavelength is 25 mm in concrete. The velocity of the wave in concrete is 4500 m/s.

Transposing the formula $v = f\lambda$ to make f the subject of the formula gives:

$$f = \frac{v}{\lambda} = \frac{4500}{0.025} = 180\,000 \text{ Hz} = 180 \text{ kHz}$$

Decibel scale

Referring again to *Fig 1*, if the amplitude of movement of the piston is increased the amplitude of movement of the air molecules will increase with a consequent increase in both the pressure variation within the wave and the energy propagated by the wave.

Since the eardrum of the human ear responds to the pressure variations in the sound wave the loudness of the sound will increase as the amplitude increases. In selecting a scale for measuring sound. It is necessary to select a method which reflects the way in which the human ear responds to the sound.

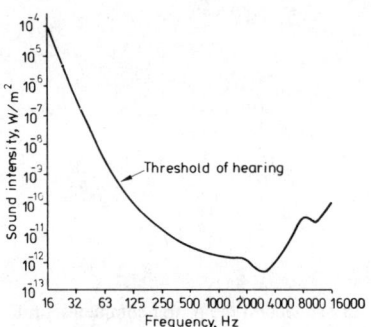

Fig 3 **Threshold of hearing**

Two important aspects need to be considered. Firstly there is a minimum sound intensity, measured in W/m^2, which the ear can detect; this is termed the threshold of hearing, and has an approximate value of 10^{-12} W/m^2 at 1000 Hz. The intensity at the threshold of hearing varies with frequency being higher at both lower and higher frequencies, this is illustrated in *Fig 3*. The range of sound power that the human ear can appreciate is truly astonishing. The most powerful sound which is likely to be encountered and can lead to pain in the ears is a factor of ten million million above the threshold.

The second aspect to consider is how the human ear responds to changes in the intensity of the sound. The Weber–Fechner law suggests that the response of the ear is logarithmic in character.

Thus the scale used to put objective numbers to sound intensities is logarithmic. A sound 10 times more powerful than another is said to have a level 10 decibels, or 10 dB, above the first sound. If it is 100 times more powerful, it is 20 dB above the first; if 1000 times more intense, it is 30 dB above the first, and so on.

The decibel difference between the two sounds can be written more formally as:

$$10 \log_{10} \frac{I_2}{I_1} \text{ dB}$$

where I_2 is the intensity of the second sound in W/m^2
and I_1 is the intensity of the first sound in W/m^2.

> *Problem 7.* A sound has 300 times the intensity of another sound. Find the decibel difference between the sounds.

In this case $I_2 = 300 \, I_1$

Thus, decibel difference $= 10 \log_{10} \dfrac{300 \, I_1}{I_1}$

$= 10 \log_{10} 300 = 24.8$ dB

> *Problem 8.* A sound has an intensity of 10^{-6} W/m^2. Determine the intensity of a sound which is 15 dB greater.

$$15 = 10 \log_{10} \frac{I_2}{10^{-6}}$$

Hence, $\log_{10} \dfrac{I_2}{10^{-6}} = 1.5$

Taking antilogarithms:

$\dfrac{I_2}{10^{-6}} = 31.6$

$I_2 = 31.6 \times 10^{-6}$
$I_2 = 3.16 \times 10^{-5}$ W/m^2

It will be noticed that the above definition does not give an absolute scale, only a means of measuring relative intensities. It needs a fixed base line from which measurements can be made. This is taken to be the threshold of hearing of 10^{-12} W/m^2.

Thus the decibel scale is defined as:

Intensity level $= 10 \log_{10} \dfrac{I}{I_o}$ dB

where I_o is the reference intensity of 10^{-12} W/m^2.

Since most of the measuring instruments used in acoustics operate on the pressure variations caused by the sound wave it is useful to define a decibel scale in terms of these pressure variations. For plane progressive waves and spherical waves at some distance from the source it can be shown that:

intensity $\propto$ (pressure)^2.

The sound pressure level (s.p.l.) is defined as

$$\text{s.p.l.} = 10 \log_{10} \frac{P^2}{P_o^2} \text{ dB}$$

where P is the root mean square pressure in the sound wave, and P_o is the reference pressure of 2×10^{-5} Pa. This reference pressure is chosen so that the intensity level and sound pressure level have the same numerical value for most practical purposes.

Problem 9. Find the s.p.l. of a sound for which the root mean square pressure is 2×10^{-3} Pa.

$$\text{s.p.l.} = 10 \log_{10} \frac{P^2}{P_o^2} \text{ dB}$$

For calculation purposes this may be more conveniently written as:

$$\text{s.p.l.} = 20 \log_{10} \frac{P}{P_o} \text{ dB}$$

$$\text{s.p.l.} = 20 \log_{10} \frac{2 \times 10^{-3}}{2 \times 10^{-5}}$$

$$= 20 \log_{10} 10^2 = 20 \times 2 = \mathbf{40 \text{ dB}}$$

ADDITION AND SUBTRACTION OF DECIBELS

Since the decibel scale is logarithmic the decibel values cannot be added or subtracted in the usual way. Only the intensities or squares of pressures can be added or subtracted. The following examples illustrate the method.

Problem 10. Two sounds of intensity level 70 dB and 75 dB are heard simultaneously. Find the resultant intensity level.

Using the definition of the intensity level and considering the first sound of 70 dB:

$$70 = 10 \log_{10} \frac{I_1}{I_o}$$

$$\log_{10} \frac{I_1}{I_o} = 7$$

By taking antilogarithms:

$$\frac{I_1}{I_o} = 1 \times 10^7$$

Similarly for the second sound of 75 dB:

$$75 = 10 \log_{10} \frac{I_2}{I_o}$$

$$\log_{10} \frac{I_2}{I_o} = 7.5; \quad \frac{I_2}{I_o} = 3.162 \times 10^7$$

If I is used to denote the total intensity then,

$$\frac{I}{I_o} = \frac{I_1}{I_o} + \frac{I_2}{I_o}$$
$$= 1 \times 10^7 + 3.162 \times 10^7$$
$$= 4.162 \times 10^7$$

The resulting intensity level is given by:

$$10 \log_{10} 4.162 \times 10^7 = \mathbf{76.2 \text{ dB}}$$

Problem 11. Two sounds of intensity level 60 dB each are produced simultaneously. Find the resultant intensity level.

For one sound:

$$60 = 10 \log_{10} \frac{I_1}{I_o}$$

$$\log_{10} \frac{I_1}{I_o} = 6$$

$$\frac{I_1}{I_o} = 1 \times 10^6$$

For two identical sounds the intensity will be doubled thus:

$$\frac{I}{I_o} = 2 \times 1 \times 10^6 = 2 \times 10^6$$

The resulting intensity level is then given by:

Intensity level = $10 \log_{10} 2 \times 10^6 = \mathbf{63 \text{ dB}}$

This illustrates a general rule that if two equal levels are added the increase is 3 dB.

Problem 12. Two sounds have sound pressure levels of 75 dB and 85 dB respectively. Find the resultant sound pressure level.

Applying the definition of sound pressure level to the first sound:

$$75 = 10 \log_{10} \frac{P_1^2}{P_o^2}$$

$$\log_{10} \frac{P_1^2}{P_o^2} = 7.5$$

$$\frac{P_1^2}{P_o^2} = 3.162 \times 10^7$$

For the second sound:

$$85 = 10 \log_{10} \frac{P_2^2}{P}$$

$$85 = 10 \log_{10} \frac{P_2^2}{P_0^2}$$

$$\log_{10} \frac{P_2^2}{P_0^2} = 8.5$$

$$\frac{P_2^2}{P_0^2} = 3.162 \times 10^8 = 31.62 \times 10^7$$

The total squared pressure P^2 is given by

$$\frac{P^2}{P_0^2} = \frac{P_1^2}{P_0^2} + \frac{P_2^2}{P_0^2}$$

$$\frac{P^2}{P_0^2} = 3.162 \times 10^7 + 31.62 \times 10^7 = 34.782 \times 10^7$$

The resulting s.p.l. is found by:

$$\text{s.p.l.} = 10 \log_{10} \frac{P^2}{P_0^2} = 10 \log 34.782 \times 10^7 = 85.4 \text{ dB}.$$

Notice that the method is identical whether intensities or pressures are used.

This example illustrates a simple rule of thumb: if two sounds differing by 10 dB or more are added there is virtually no increase in level above the highest value.

Problem 13. In a factory six similar machines running together produce a sound pressure level of 90 dB. Find the sound pressure level if four of the machines are turned off.

For all six machines:

$$90 = 10 \log_{10} \frac{P^2}{P_0^2}$$

$$\log_{10} \frac{P^2}{P_0^2} = 9$$

$$\frac{P^2}{P_0^2} = 1 \times 10^9$$

For one machine the value of $\frac{P^2}{P_0^2}$ would be: $\frac{1 \times 10^9}{6}$

For two machines left running:

$$\frac{P^2}{P_0^2} = \frac{2 \times 1 \times 10^9}{6} = 0.3333 \times 10^9 = 3.3333 \times 10^8$$

Thus the s.p.l. is found as:

$$\text{s.p.l.} = 10 \log_{10} \frac{P^2}{P_0^2} = 10 \log_{10} 3.333 \times 10^8 = 85.2 \text{ dB}.$$

Sound insulation

Sound insulation is concerned with the ability of partitions for example walls, floors and roofs, to reduce the transmission of sound from the room or building in which it is produced to the room or building in which it is received.

It is usual to divide sound insulation into two classifications. Firstly airborne sound

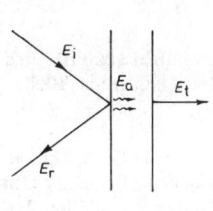

Fig 4 Sound transmission through a partition

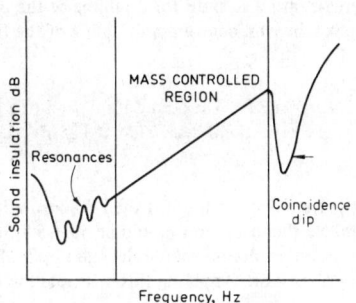

Fig 5 Variation of sound insulation with frequency

insulation where the sound source is in the air and the receiver, usually the human ear, is also in air. A simple example is the transmission of the sound from a television set through a partition wall into another room or house. Secondly impact sound insulation where the sound originates from an impact on the construction and is received by a receiver in air. A simple example is the sound created by footsteps on a floor creating noise received in the room below.

AIRBORNE SOUND INSULATION

It is necessary to consider the transmission of sound through a partition. *Fig 4* shows a sound wave incident on a partition having energy E_i. Part of the energy is reflected by the surface and part is absorbed, these are shown as E_r and E_a respectively. Since the sound wave sets the material of the partition into vibration a proportion of the energy is transmitted to the air on the remote side of the partition. The energy transmitted is denoted as E_t in *Fig 4*.

The proportion of the incident energy which is absorbed is called the absorption coefficient and is mainly a property of the surface and may well reach 90% for sound absorbing materials. The absorbed energy will be partially converted to heat and partially transmitted.

The sound reduction index is used to specify the sound insulation and represents the difference in sound pressure level between the source room and the receiving room when measured and calculated in accordance with the appropriate British Standard.

FACTORS AFFECTING THE SOUND INSULATION OF SINGLE LEAF PARTITIONS

Since the incident sound sets the material of the partition into vibration the amplitude of this vibration will depend on the mass per unit area of the partition. The greater the mass per unit area the less the vibration and hence less sound will be transmitted. It is thus expected that the sound insulation will improve as the mass per unit area of the partition increases. This is called the **Mass Law**.

The sound insulation also depends on frequency as illustrated in *Fig 5*. For many practical partitions over the frequency range 100 Hz to 3150 Hz the mass controlled region is the most important and it will be seen that, in this region, the sound insulation increases with frequency. From a practical point of view the insulation is likely to increase, at most, 5 dB for doubling of the frequency. This is often stated as 5 dB per octave, since an octave is a doubling of the frequency.

> *Problem 14.* The sound insulation of a brick wall is 45 dB at 1000 Hz, predict the likely sound insulation at 125 Hz, 250 Hz, 500 Hz, 2000 Hz and 4000 Hz.

If the frequency is doubled then the insulation should increase by 5 dB. Thus at 2000 Hz the insulation should be 45 + 5 = 50 dB. Similarly if the frequency is halved the sound insulation should decrease by 5 dB, thus at 500 Hz the insulation should be 45 − 5 = 40 dB. Applying this principle the following values are obtained.

Frequency (Hz)	125	250	500	1000	2000	4000
Insulation (dB)	30	35	40	45	50	55

AVERAGE SOUND INSULATION

The sound insulation is often quoted as a single figure which is the average performance over the frequency range 100 Hz to 3150 Hz. This figure for the average sound insulation is very nearly equal to its performance at 500 Hz. The average sound insulation, for a single leaf partition, may be approximately predicted by the equation:

$R = 10 + 14.5 \log_{10} m$

where R is the average sound reduction in dB and m is the mass per unit area of the partition in kg/m^2.

> *Problem 15.* Predict the average sound insulation that will be provided by a one-brick wall having a mass per unit area of 415 kg/m^2.

Using the formula:

$R = 10 + 14.5 \log_{10} m$
$R = 10 + 14.5 \log_{10} 415 = 48$ dB

> *Problem 16.* For the brick wall in *Problem 15* predict the likely sound insulation at 125, 250, 500, 1000, 2000 and 4000 Hz.

In *Problem 15* the average sound insulation was found to be 48 dB and this may be taken to be its performance at 500 Hz. It is probable that the insulation will increase

5 dB each time the frequency is doubled. The following table is then readily constructed.

Frequency (Hz)	125	250	500	1000	2000	4000
Insulation (dB)	38	43	48	53	58	63

Factors affecting sound insulation in practice

FLANKING TRANSMISSION

Fig 6 shows some of the paths by which sound can travel from the source room to the receiving room. It will be seen that sound can be transmitted by numerous paths; the sound reduction index applies to the direct path labelled A in the diagram. The other paths show flanking transmission and the amount of sound transmitted along these paths will depend on the properties of the side walls. The more massive the side walls the less sound will be transmitted by them. Flanking transmission is usually not important for partitions with an insulation value below about 35 dB but is of the utmost importance when the insulation of the partition reaches 50 dB.

Fig 6 Flanking transmission

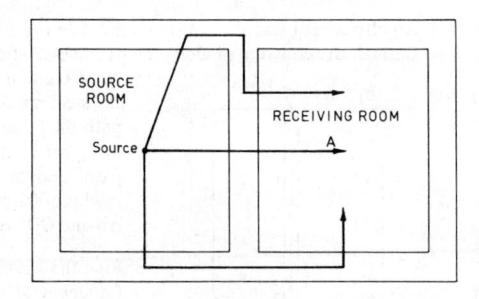

COMPOSITE PARTITIONS

The existence of an area of low sound insulation in a partition of otherwise high sound insulation value will dramatically reduce the overall insulation of the partition. For instance a door having a sound insulation of 20 dB in a plastered brick wall of sound insulation 45 dB is likely to reduce the overall sound insulation of the partition to 30 dB. The actual reduction depends on the areas of the door and the brickwork.

COMPLETENESS

Air gaps in the partition will reduce the sound insulation considerably. Many possible causes of gaps exist. The following are just a few:
gaps around doors and windows;
gaps around service pipes and cables;
air spaces through partitions above suspended ceilings.

IMPACT SOUND INSULATION

The noise source, in this case, is impacts upon the structure. A simple instance is that of footsteps on a floor. A floor with a good airborne sound insulation may provide very little impact sound insulation (a solid concrete floor is a good example). Although the vibration of the floor due to impact is reduced if the mass of the floor is increased by far the best way of improving the impact sound insulation is to introduce a resilient layer to absorb the energy of the impact. This may be achieved in practice either by the use of resilient floor coverings or by the use of a floating floor construction.

Exercises (*Answers on page 106*)

1 Calculate the velocity of sound in air at a temperature of 15° C.

2 Find the time taken for an ultrasonic pulse to pass through a 5 m length of concrete if the velocity of the pulse in concrete is 5000 m/s.

2 In a hall with a public address system a member of the audience is 17 m from the lecturer. The lecturer is using the public address system and the person is 3 m from the nearest loudspeaker. Assuming that the electrical signal from the microphone reaches the loudspeaker instantaneously, find the delay between the arrival of the sound from the loudspeaker and that from the lecturer. Assume the velocity of sound in air to be 340 m/s.

4 An ultrasonic pulse starts at the point A in *Fig 7* and can travel to D either directly through the concrete or along the path ABCD. For the path ABCD the parts AB and CD are in concrete and the part BC is along the steel reinforcement. Determine by which path the pulse first reaches the point D and the delay between the arrivals by the two different paths. Assume that the velocity of the pulse in steel is 6000 m/s and in concrete is 4500 m/s. AB and CD are at 45 °C to AD

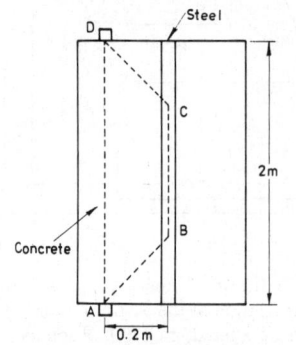

Fig 7 Diagram for Exercise 4

5 Find the wavelength of a sound having a frequency of 125 Hz if the velocity of sound in air is 340 m/s.

6 The velocity of an ultrasonic pulse in concrete is found to be 5000 m/s. Two ultrasonic transducers are placed 250 mm either side of a crack in a concrete beam and the time for the pulse to be transmitted was found to be 105 μs. Determine the depth of the crack.

7 Standing waves can be created between two parallel walls in a room if the distance between the walls is $\lambda/2, \lambda, 3\lambda/2\ldots$, where λ is the wavelength of sound in air. For two walls separated by a distance of 3 m calculate the frequencies of the first five possible standing wave systems, assuming the velocity of sound to be 340 m/s.

8 A sound has one million times the intensity of another sound. Determine the decibel difference between the sounds.

9 Two sounds have intensities of 3×10^{-6} W/m^2 and 7×10^{-5} W/m^2. Determine the decibel difference between these sounds.

10 Find the increase in intensity level if the intensity of a sound is doubled.

11 Find the increase in sound pressure level if the sound pressure is doubled.

12 Three sounds each have a sound pressure level of 65 dB. Calculate the resultant sound pressure level.

13 In determining the predicted noise level due traffic on a proposed new road it was necessary to add the following sound pressure levels: 68 dB, 72 dB and 74 dB. Calculate the total sound pressure level.

14 In a machine shop the sound pressure level due to 10 identical machines is 94 dB. How many machines must be stopped in order to reduce the sound pressure level to 90 dB.

15 The background sound pressure level in a classroom is 50 dB. With 10 students working in the room the sound pressure level rises to 60 dB. Find the likely sound pressure level with 30 students in the room.

16 A wall has a sound insulation of 50 dB at 1000 Hz. Predict the likely sound insulation at 125, 250, 500, 2000 and 4000 Hz, assuming that the sound insulation increases 5 dB per octave.

17 The wall in Exercise 16 separates a busy road from a lecture room. The sound pressure levels, measured just outside the wall at a number of frequencies, are given in the table below.

Frequency (Hz)	125	250	500	1000	2000	4000
s.p.l. (dB)	80	74	73	73	71	66

Determine the probable sound levels inside the room at each of the frequencies.

18 The intensity of the sound incident upon a partition is 10^{-3} W/m^2. The sound reduction index of the partition is 46 dB. Determine the intensity of the sound on the receiving side of the partition.

19 Predict the likely average sound insulation of a 110 mm brick wall. The density of the brickwork may be taken as 1650 kg/m^3.

20 Repeat Exercise 19 for 220 mm and 330 mm thick brick walls. State the increase in sound insulation achieved by doubling the mass per unit area of a wall.

21 The material used for the construction of an acoustic screen is required to have a sound reduction index of at least 30 dB. Suggest a minimum mass per unit area of the material of the screen.

22 Of the sound intensity incident upon a partition 5% is reflected, 92% is absorbed and the remainder is transmitted. Find the sound reduction index for the partition.

23 Two sounds each have a sound pressure level of 58 dB. Select from the options below the resultant sound pressure level.
(a) 116 dB; (b) 63 dB; (c) 61 dB; (d) 58 dB.

24 Two sounds have sound pressure levels of 60 dB and 82 dB. Select from the options below the resulting sound pressure level, to the nearest decibel.
(a) 142 dB; (b) 85 dB; (c) 63 dB; (d) 82 dB.

25 Six machines produce a sound pressure level of 80 dB. Select from the options below the resulting sound pressure level when three of the machines are stopped.
(a) 77 dB; (b) 83 dB; (c) 74 dB; (d) 40 dB.

26 The sound pressure level on the receiving side of a partition is 53 dB. The sound reduction index of the partition is 27 dB. Select from the options below the sound pressure level on the source side.
(a) 80 dB; (b) 53 dB; (c) 26 dB; (d) 42 dB.

27 The sound insulation of a partition is 35 dB at 500 Hz. Select from the options below the likely sound insulation at 2000 Hz.
(a) 35 dB; (b) 25 dB; (c) 45 dB; (d) 40 dB.

28 The mass per unit area of a partition is doubled. Select from the options below the change in the average sound insulation in decibels.
(a) it will increase by four to five decibels;
(b) it will double in value;
(c) it will halve in value;
(d) it will decrease by four to five decibels.

29 A partition wall was to be constructed in 100 mm lightweight aerated blocks plastered on both sides. Instead it was constructed in 110 mm wire cut bricks plastered on both sides. The sound insulation was found to be much lower than expected. Delete the incorrect options from the following possible causes.
(a) the partition had only been built up to a plasterboard ceiling with a large air-space above it.
(b) the materials used were too heavy.
(c) the gaps around the service pipes were not made good.
(d) the bricks were laid frog down instead of frog up thus reducing the weight of the partition.

3 Applied mechanics

Stress

When an external force is applied to a body internal forces are mobilised to resist deformation and the material is said to be in a state of stress.

The intensity of stress over any plane in the body is defined by:

$$\text{Stress} = \frac{\text{Force}}{\text{Area}}$$

The units of stress are those of force divided by those of area; for example N/m^2, kN/m^2, N/mm^2 and MN/m^2. Note that $1\ N/mm^2 = 1\ MN/m^2$.

A number of types of stress exist. *Fig 1(a)* indicates direct tension. The normal tensile stress acting on a typical section, ABCD, at right angles to the direction of the force is found by:

$$\text{Normal tensile stress} = \frac{\text{Force}}{\text{Area ABCD}}$$

or, in symbols

$$\sigma = \frac{F}{A}$$

Fig 1(b) shows the case of direct compression and the normal compressive stress is again the force divided by the area ABCD. This is similarly calculated using the formula

$$\sigma = \frac{F}{A}$$

Note that σ is conventionally used to denote normal stresses.

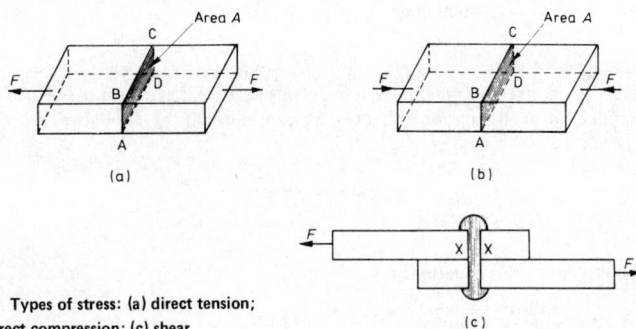

Fig 1 Types of stress: (a) direct tension; (b) direct compression; (c) shear

Fig 1(c) shows shearing forces acting on a rivet. The shear stress is found by dividing the force by the cross sectional area of the rivet along the section X–X. It is conventional to use the symbol τ to denote shear stress hence we can write

$\tau = \dfrac{F}{A}$, where A is the cross sectional area.

Problem 1. A steel bar of cross section 20 mm × 30 mm is subject to an axial pull of 60 kN. Determine the normal tensile stress.

$$\text{stress} = \dfrac{\text{force}}{\text{area}}$$

$$= \dfrac{60 \times 1000}{20 \times 30} = 100 \text{ N/mm}^2$$

Problem 2. A concrete cube of side 150 mm is tested in compression. The failing load was 700 kN. Calculate the compressive stress at failure.

$$\text{stress} = \dfrac{\text{force}}{\text{area}}$$

$$= \dfrac{700 \times 1000}{150 \times 150} = 31.1 \text{ N/mm}^2$$

Problem 3. Determine the cross sectional area of a tie rod which is to be subjected to an axial load of 75 kN if the normal tensile stress is not to exceed 125 N/mm².

$$\sigma = \dfrac{F}{A}$$

Thus by rearranging this equation:

$$A = \dfrac{F}{\sigma} = \dfrac{75 \times 1000}{125} = 600 \text{ mm}^2$$

Problem 4. A rivet, 20 mm in diameter, is in single shear as shown in *Fig 1(c)*. Determine the maximum force that can be transmitted if the shear stress is not to exceed 100 N/mm².

$$\tau = \dfrac{F}{A}$$

By rearrangement of this equation:

$$F = A \times \tau = \dfrac{\pi (20)^2}{4} \times 100 = 31420 \text{ N} = 31.42 \text{ kN}$$

Strain

When an external force is applied to a body deformation occurs. The change of shape is measured by the strain which for direct compression or direct tension is defined by:

$$\text{strain} = \frac{\text{change in length}}{\text{original length}}$$

or, in symbols,

$$\epsilon = \frac{\delta l}{l}$$

where ϵ = strain;

δl = change in length. Note that this is to be treated as a single symbol.
l = original length.

The strain is dimensionless since it is a length divided by a length; it is necessary to ensure that both the change in length and original length are measured in the same units.

Problem 5. A steel rod 1.5 m long was subjected to an axial tensile force of 130 kN and extended by 2.0 mm. Calculate the strain.

$$\text{strain} = \frac{\text{change in length}}{\text{original length}} = \frac{2.0}{1.5 \times 1000} = 0.00133$$

Problem 6. A metal rod is subjected to axial tension which causes a strain of 0.08%. If the original length of the bar is 0.5 m find the change in length of the bar.

Note that the strain is quoted as a percentage so that the actual strain is 0.08/100 = 0.0008.

Using the given formula: $\epsilon = \frac{\delta l}{l}$ transposition of the formula gives:

$\delta l = \epsilon l$
$ = 0.0008 \times 0.5$ m
$ = 0.0008 \times 0.5 \times 1000$ mm = 0.4 mm
Change in length = 0.4 mm

Modulus of elasticity

All materials deform when subjected to stress. For many materials when the stress is removed the material returns to its original shape, such materials are termed **elastic**. For all materials there exists a value of the stress above which they do not return entirely to their original shape when the stress is removed. This value of the stress is the **elastic limit**. Many materials used in construction are elastic and it is useful to study the relationship between stress and strain.

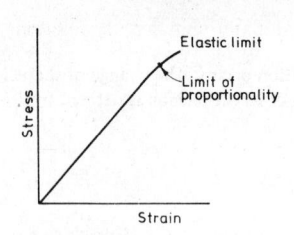

Fig 2 shows graphically the behaviour of an elastic material, subjected to direct stress up to the elastic limit. It will be noted that this is a linear relationship up to the limit of proportionality. This linear relationship is so important that it gave rise to the well known **Hooke's Law**, named after the original discoverer. Hooke's Law states that the strain is proportional to the stress up to the limit of proportionality.

Fig 2 Behaviour of an elastic material up to the elastic limit

Straight line relationships are very important in technology since they are readily specified and lead to easy prediction of behaviour. In this case, since the line passes through the origin it is completely specified by its gradient or slope. The gradient of the line is a characteristic property of the material and is the ratio of stress to strain. This ratio is known as the **Modulus of Elasticity** or **Young's Modulus**. It is denoted by the symbol E.

$$\text{Modulus of Elasticity} = \frac{\text{Direct Stress}}{\text{Direct Strain}}$$

Note that this modulus refers to direct stress and direct strain, other moduli exist for other types of stress and strain.

UNITS OF YOUNG'S MODULUS

Since strain is a dimensionless quantity the units of Young's modulus are those of stress. For many materials the numerical value of Young's modulus is very large and it is common to use either kN/mm^2 or GN/m^2 (10^9 N/m^2).

Problem 7. In a tensile test on a 10 mm diameter steel bar a load of 20 kN caused an extension of 0.061 mm measured on a length of 50 mm of the bar. Assuming that the limit of proportionality had not been reached calculate the value of Young's modulus.

Cross sectional area of the bar = $\pi.10^2/4$ = 78.54 mm^2.

$$\text{Stress} = \frac{\text{load}}{\text{area}} = \frac{20}{78.54} = 0.2546 \text{ kN/mm}^2$$

$$\text{Strain} = \frac{\text{change in length}}{\text{original length}} = \frac{0.061}{50} = 0.00122$$

$$\text{Young's modulus}, E = \frac{\text{stress}}{\text{strain}} = \frac{0.2546}{0.00122} = 208.7 \text{ kN/mm}^2$$

Problem 8. A timber tie 3 m long is subjected to a stress of 8 N/mm^2. Calculate the elongation of the tie if Young's modulus for the timber is 11 kN/mm^2.

In answering this question extreme care is necessary with the units, and the following should be noted.

Stress = 8 N/mm^2
E = 11 kN/mm^2 = 11000 N/mm^2
length = 3 m = 3000 mm

By transposing the formula $E = \dfrac{\text{stress}}{\text{strain}}$

the strain can be written as:

$$\text{strain} = \frac{\text{stress}}{E} = \frac{8}{11000} = 0.0007273$$

From the definition of strain:

change in length = strain × original length = 0.0007273 × 3000 = **2.18 mm**

Problem 9. In a test on a machine a strain gauge was attached to a steel column having a diameter of 80 mm. When a tensile stress was applied the gauge indicated a strain of 189×10^{-6}. Assuming Young's modulus for the steel to be 210 kN/mm^2 determine the applied load.

From the definition of Young's modulus:

stress = E × strain = $210 \times 189 \times 10^{-6}$ kN/mm^2 = 3.969×10^{-2} kN/mm^2

From the definition of stress

$$\text{load} = \text{stress} \times \text{area} = 3.969 \times 10^{-2} \times \pi \times \frac{80^2}{4} \text{ kN} = \mathbf{199.5 \text{ kN}}$$

Tensile strength

The complete load-extension curve for mild steel is shown in *Fig 3*. It is observed that the load reaches a maximum and then decreases until the failure point is reached. The

Fig 3 Load-extension curve for mild steel

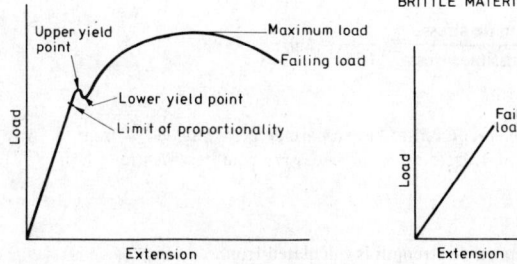

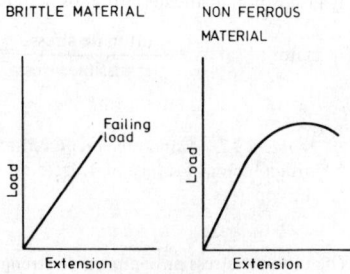

Fig 4 Load-extension curves for a brittle material and a non ferrous metal

tensile strength, sometimes referred to as the ultimate tensile stress, is defined as:

$$\text{tensile strength} = \frac{\text{maximum load}}{\text{original cross sectional area}}$$

The above definition of tensile strength applies to any material subjected to a tensile test. The shape of the load extension curve will depend on the material and a few examples are shown in *Fig 4*.

> *Problem 10.* In a tensile test on a steel bar of 10 mm diameter the maximum load was 35 kN. Calculate the tensile strength of the steel.

$$\text{tensile strength} = \frac{\text{maximum load}}{\text{original cross sectional area}}$$

$$= \frac{35 \times 1000}{\pi \times \frac{10^2}{4}} = 446 \text{ N/mm}^2$$

Factor of safety

The working stress to which a member is subjected must not reach the ultimate stress for the material otherwise failure will occur. The maximum permissible value of the working stress is defined by:

$$\text{permissible stress} = \frac{\text{ultimate stress}}{\text{factor of safety}}$$

The factor of safety used varies between 1.5 and 5 depending on the material and the application.

> *Problem 11.* The tensile strength of a bar is 450 N/mm². If the maximum permissible stress is 155 N/mm² calculate the factor of safety.

Transposition of the definition of permissible stress gives:

$$\text{factor of safety} = \frac{\text{ultimate stress}}{\text{permissible stress}} = \frac{450}{155} = 2.9$$

> *Problem 12.* Concrete is required to have a permissible stress of 7 N/mm² with a factor of safety of 4. Determine the necessary compressive strength of the concrete.

The ultimate stress or compressive strength is calculated from:

compressive strength = permissible stress × factor of safety
$$= 7 \times 4 = 28 \text{ N/mm}^2$$

It is to be noted that the factor of safety as defined above is not the only way of ensuring the safety of components of structures. At a later stage the student will encounter a number of other methods including yield factor and load factor.

Beams

MOMENT OF A FORCE

The turning effect of a force about a point is measured by the **Moment of the Force**. The moment is defined as:

moment = force × perpendicular distance of its line of action from the point.

Fig 5 illustrates how the perpendicular distance is obtained. In each case:
moment = $F \times d$.

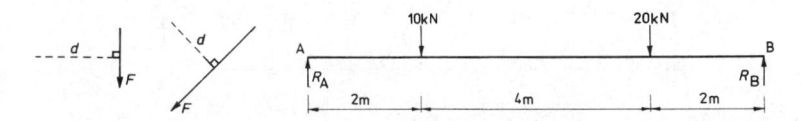

Fig 5 Moment of a force   **Fig 6 Loaded beam for the calculation of reactions**

BEAM REACTIONS

Consider the beam shown in *Fig 6* subjected to vertical loading. It is necessary to find the support reactions R_A and R_B which will be vertical since the loading is vertical. To obtain the values of the reactions the following conditions of equilibrium are used:
(i) The algebraic sum of the vertical forces is zero. This may be expressed simply as:

sum of forces upwards = sum of forces downwards

(ii) The algebraic sum of the moments of the forces about any point is zero. This may be alternatively written as:

sum of anticlockwise moments = sum of the clockwise moments.

Applying the first condition to the beam in *Fig 6*:

$R_A + R_B = 10 + 20 = 30$ kN

In applying the second condition it is possible to take moments about any point but in order to simplify the calculation it is convenient to take moments about the reaction points A and B.

Taking moments about A:

$8R_B = 10 \times 2 + 20 \times 6 = 140$
$R_B = 17.5$ kN

Note that since R_A passes through A it has no moment about A. The reader is advised to check very carefully the distances between A and each of the loads.

Since the sum of R_A and R_B is known to be 30 kN it would be easy to find R_A by subtraction but it is better practice to take moments about B in order to find R_A and check that R_A and R_B together give 30 kN. This will ensure that the reaction values are correct before proceeding with any further calculations. Taking moments about B:

$10 \times 6 + 20 \times 2 = 8R_A$
$8R_A = 100$
$R_A = 12.5$ kN

Now, $R_A + R_B = 17.5 + 12.5 = 30$ kN

This is correct total, thus ensuring that the values of R_A and R_B are in themselves correct.

Problem 13. Calculate the support reactions for the beam loaded as shown in *Fig 7.*

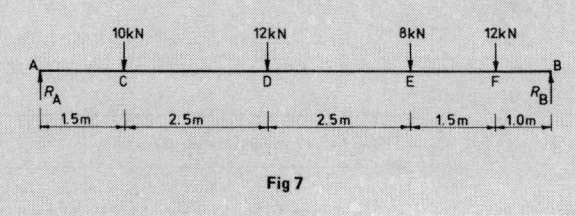

Fig 7

Applying the first condition: $R_A + R_B = 10 + 12 + 8 + 12 = 42$ kN
Taking moments about A:

$9R_B = 10 \times 1.5 + 12 \times 4 + 8 \times 6.5 + 12 \times 8$
$9R_B = 15 + 48 + 52 + 96$
$9R_B = 211$
$R_B = 211/9 = 23.44$ kN

Taking moments about B:

$12 \times 1 + 8 \times 2.5 + 12 \times 5 + 10 \times 7.5 = 9R_A$
$12 + 20 + 60 + 75 = 9R_A$
$9R_A = 167$
$R_A = 167/9 = 18.56$ kN

Check: $R_A + R_B = 23.44 + 18.56 = 42$ kN

Again the reader is advised to check very carefully the distances from A when moments are taken about A and similarly for point B.

SHEAR FORCE

Consider the section at X–X on the beam illustrated in *Fig 8*. To the left of this section there exists a net vertical force given by $12.5 - 10 = 2.5$ kN vertically upwards. Thus there is a tendency for the left-hand section to move vertically upwards.

To the right of the section X–X the net vertical force is $20 - 17.5 = 2.5$ kN vertically downwards. Hence the right hand section of the beam tends to move vertically downwards. These equal and opposite forces constitute a shear force across the section X–X.

If the beam is to remain in equilibrium this shear force must be resisted by the shear strength properties of the material. The shear force is formally defined as the algebraic sum of all the forces acting on one side of the section.

Since force is a vector quantity it is necessary to specify the direction of the force as well as its magnitude and a sign convention is necessary. The following simple definition is adequate at this stage to ensure that the usual sign convention is followed:

Shear Force = Reaction − Sum of loads to the left of the section

If the result of this calculation is positive the left-hand section of the beam is tending to move upwards and if it is negative it is tending to move downwards. A shear force diagram illustrates the variation of shear force along the length of the beam.

Problem 14. Draw the shear force diagram for the beam shown in *Fig 8*.

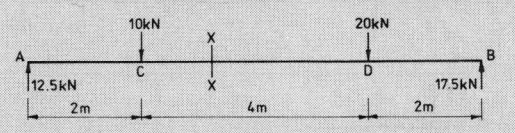

Fig 8 Loaded beam for calculation of shear force

For any section between A and C the forces acting to the left of the section are:

reaction = 12.5 kN
sum of loads to left of section = 0
shear force = 12.5 − 0 = 12.5 kN

For any section between C and D the forces acting to the left of the section are:

reaction = 12.5 kN
sum of loads to left of section = 10 kN
shear force = 12.5 − 10 = 2.5 kN

For any section between D and B the forces acting to the left of the section are:

reaction = 12.5 kN
sum of loads to left of section = 10 + 20 = 30 kN
shear force = 12.5 − 30 = − 17.5 kN

Thus the shear force takes the following values:

between A and C
  12.5 kN
between C and D
  2.5 kN
between D and B
  −17.5 kN

The shear force diagram is shown in *Fig 9*.

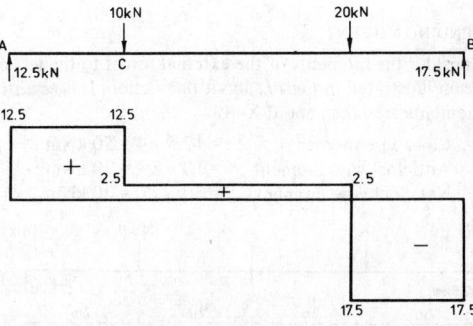

Fig 9 Loaded beam and shear force diagram for Problem 14

Problem 15. Draw the shear force diagram for the beam loaded as shown in *Fig 7*.

From *Problem 13* it is known that R_A = 18.56 kN and R_B = 23.44 kN.
The values of the shear force are:

Between A and C:	18.56 − 0	= 18.56 kN
Between C and D:	18.56 − 10	= 8.56 kN
Between D and E:	18.56 − (10 + 12)	= −3.44 kN
Between E and F:	18.56 − (10 + 12 + 8)	= −11.44 kN
Between F and B:	18.56 − (10 + 12 + 8 + 12)	= −23.44 kN

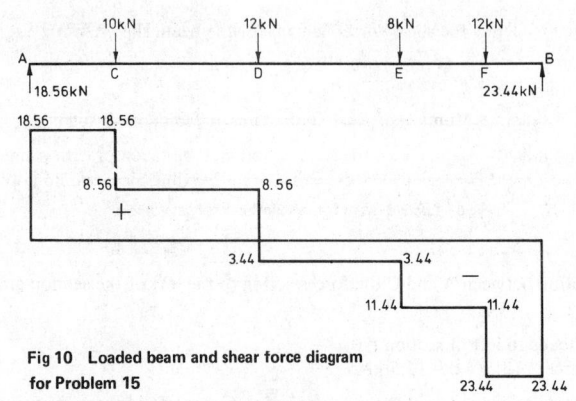

Fig 10 Loaded beam and shear force diagram for Problem 15

The shear force diagram is shown in *Fig 10*.
The following points should now be clear:
(i) for a simply supported beam carrying point loads only the shear force diagram consist of a series of horizontal straight lines.
(ii) steps occur in the shear force diagram at each of the point loads. The magnitude of the step is equal to the value of the point load.
(iii) the value of the shear force near the reactions is numerically equal to the reaction values. This provides a check on the construction of the shear force diagram.

BENDING MOMENT

Consider the moments of the external forces to the left of the section X–X, on the beam illustrated in *Fig 11*, about the section. It is essential to appreciate that the moments are taken about X–X.

Clockwise moment	= 12.5 × 4 = 50 kNm
Anticlockwise moment	= 10 × 2 = 20 kNm
Net clockwise moment	= 50 − 20 = 30 kNm

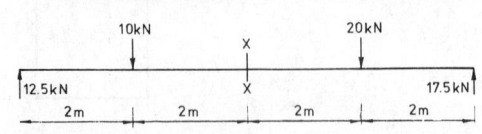

Fig 11 Loaded beam for calculation of bending moment

This net moment of the external forces tends to cause rotation in a clockwise direction of the left hand section of the beam. This is manifested by the bending of

42

the beam. Similarly taking moments about X–X for the forces to the right of the section:

Anticlockwise moment = 17.5 × 4 = 70 kNm
Clockwise moment = 20 × 2 = 40 kNm
Net anticlockwise moment = 70 − 40 = 30 kNm

Thus the right hand section tends to rotate in an anticlockwise direction.

The net external moment at a section is termed the bending moment. The formal definition of the bending moment at a section is the algebraic sum of the moments about the section of the forces on one side of the section.

A sign convention is necessary and it is usual to assume the bending moment to be positive if it causes sagging, as for the beam considered above, and negative if it causes hogging.

For the simply supported beams considered in this chapter the following simple rule ensures the correct sign convention:

Bending Moment = Moment of Reaction − Sum of moments of loads

The moments must be taken about the section and may be taken on either side of it. The bending moment diagram shows the variation of bending moment along the length of the beam.

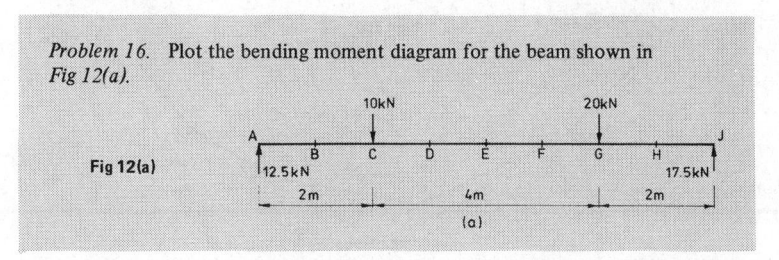

Problem 16. Plot the bending moment diagram for the beam shown in Fig 12(a).

Fig 12(a)

The bending moments will be calculated at the points A to J which are at 1 m intervals along the beam. By considering moments of the forces to the left of the section the results can be tabulated as follows.

Section	Bending Moment (kNm)	
A	0 (since reaction has no moment at A)	
B	12.5 × 1	= 12.5
C	12.5 × 2 − 10 × 0	= 25
D	12.5 × 3 − 10 × 1	= 27.5
E	12.5 × 4 − 10 × 2	= 30
F	12.5 × 5 − 10 × 3	= 32.5
G	12.5 × 6 − 10 × 4 − 20 × 0	= 35
H	12.5 × 7 − 10 × 5 − 20 × 1	= 17.5
J	12.5 × 8 − 10 × 6 − 20 × 2	= 0

These values are shown in *Fig 12(b)*. The convention used here for constructing the diagram is that the values are plotted on the tension side of the beam.

The following points should be observed from the above example:
(i) the bending moment diagram consists of a series of straight lines when only point loads are considered.

Fig 12(b)

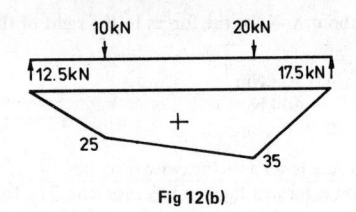

(ii) the straight lines change direction only at the point loads; thus the bending moment need only be calculated at the point loads.
(iii) if the beam is simply supported at its ends then the bending moment at the supports is zero.
(iv) bending moments may be taken either to the left or right of the section.

Problem 17. Plot the bending moment diagram for the beam shown in *Fig 13(a)*.

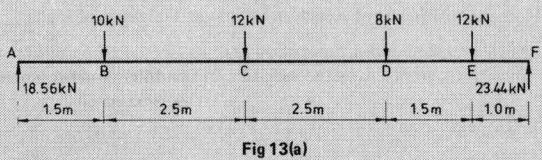

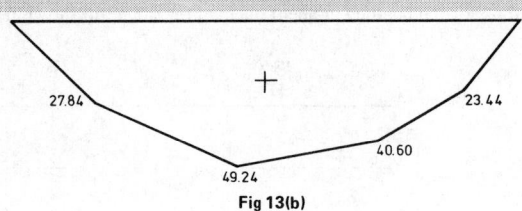

Fig 13(a)

Fig 13(b)

Section	Bending Moment (kNm)		
A	0		
B	18.56×1.5	$= 27.84$	moments taken
C	$18.56 \times 4 - 10 \times 2.5$	$= 49.24$	to left of section
D	$23.44 \times 2.5 - 12 \times 1.5$	$= 40.6$	moments taken
E	23.44×1	$= 23.44$	to right of section
F	0		

The bending moment diagram is shown in *Fig 13(b)*.

Problem 18. For the beam loaded as shown in *Fig 14(a)* calculate the reactions and construct the shear force and bending moment diagrams.

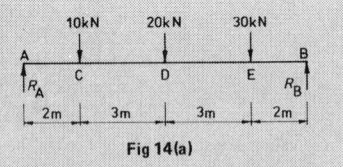

Fig 14(a)

Calculation of the reactions must be performed first.

$R_A + R_B = 60$ kN.

44

Moments about A:

$10R_B = 10 \times 2 + 20 \times 5 + 30 \times 8$

$10R_B = 360$ kN

$R_B = 36$ kN

Moments about B:

$10 \times 8 + 20 \times 5 + 30 \times 2 = 10R_A$

$10R_A = 240$

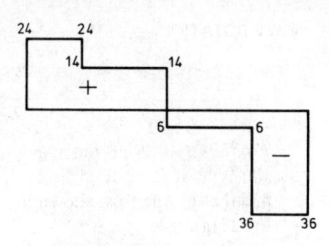

$R_A = 24$ kN

Fig 14(b) Shear force diagram

The shear force diagram is shown in *Fig 14(b)*.

The following values of bending moment are necessary in order to plot the bending moment diagram.

Section	Bending Moment (kNm)	
C	24×2	= 48
D	$24 \times 5 - 10 \times 3$	= 90
E	36×2	= 72

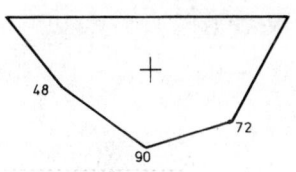

Fig 14(c) Bending moment diagram for Problem 18

The bending moment diagram is illustrated in *Fig 14(c)*. By comparing the bending moment and shear force diagrams it will be seen that the maximum bending moment occurs at the zero of the shear force. This is generally true for all systems of loading of beams.

Internal forces in structures

Any structure may be regarded as designed to transmit forces from their point of application to the points of support. A beam transmits applied loads to the supporting walls or columns. A roof truss transmits the loads due to the roofing material, the wind loads and snow loads to the supports of the truss.

For the complete structure to be in equilibrium the external forces must be in equilibrium. This concept is clearly seen in calculating beam reactions. The external forces create internal forces within the structure and each point within the structure must be in equilibrium under the action of the internal and external forces.

Internal forces in a truss

To simplify the analysis of a framed structure it is useful to assume that the weight of the members is negligible and that the joints are pin jointed so that each of the members is either subjected to purely tensile forces or purely compressive forces. In designing a truss it is first necessary to find the magnitude and nature of the forces in each of the members.

One method of doing this is to realise that each joint is in equilibrium under the action of the internal and external forces. It is well known that if a number of coplanar forces acting at a point are in equilibrium then a closed polygon may be drawn whose sides, taken in order, represent the forces in both magnitude and direction. Hence all that is necessary is to construct the polygon of forces for each joint. The use of Bow's notation and a systematic approach to the order in which the polygons are drawn will facilitate a rapid solution.

BOW'S NOTATION

Problem 19. Find the magnitude and type of force acting in each member of the truss shown in Fig 15(a).

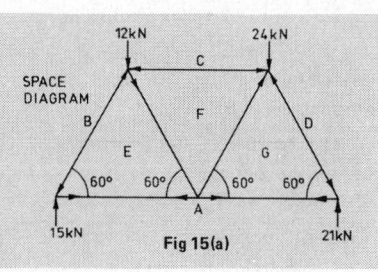

Fig 15(a)

Fig 15(b)

Step 1 Letter the spaces between the forces with capital letters. It is helpful to proceed clockwise, lettering the spaces between external forces first.

Step 2 Draw to a suitable scale the load line such that:
'ab' represents a force of 15 kN upwards
'bc' represents a force of 12 kN downwards
'cd' represents a force of 24 kN downwards
'da' represents a force of 21 kN upwards.
These are drawn vertically, starting at the left-hand support, since the forces are vertical. Check this as shown in *Fig 15(b)*.

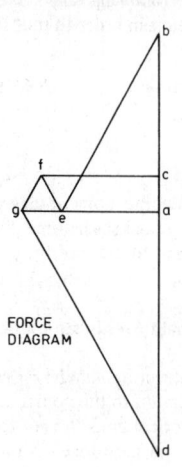

Step 3 Starting at the left hand support construct a triangle of force by drawing 'be' parallel BE and 'ea' parallel to EA. Always proceed in the same direction around the joint — in this case *clockwise*. The point 'e' on the force diagram is located as shown. At this stage the arrow-heads showing the direction of the forces are placed near the joint. On the force diagram the line 'be' is inclined at 60° downwards and to the left therefore the force in BE is parallel to it and is downwards and to the left. Insert an arrow head near the joint downwards and to the left. Similarly 'ea' on the force diagram is to the right so an arrow head is put on EA to the right near the joint. Note that all notation is used in the clockwise sense.

Step 4 Proceed across the truss to the right to locate the next joint to be considered. Draw on the force diagram 'cf' parallel to CF, that is horizontal and 'fe' parallel to FE thus locating the point 'f'. Note that 'eb' is already drawn. Insert arrow heads as previously described.

Step 5 Consider the next joint encountered on moving to the right; this is the one at the bottom centre of the truss. Draw 'fg' on force diagram parallel to FG and draw 'ga' parallel to GA; this locates the point 'g'. Note that 'ae' and 'ef' have already been drawn. Insert arrowheads as before.

46

Step 6 Proceed to the next joint. Draw 'gd' on force diagram parallel to GD. 'gf' and 'cf' already exist. Insert arrow heads near the joint.

Step 7 Considering the last joint the lines 'ag' and 'gd' are already drawn on the force diagram. Insert arrow heads.

Step 8 The magnitudes of the forces in the members can now be obtained by scaling the appropriate lines in the force diagram. From the arrowheads inserted on the space diagram the nature of the force in the member can be found. If the arrow heads on a member are pointing towards each other the member is pulling inwards to resist tension forces applied to it; the member is in tension and is a tie. Conversely if the arrow heads on a member are pointing away from each other the member is resisting compression forces applied to it and is thus in compression and is a strut.

The magnitude and nature of the forces are as follows.

BE	17.36 kN	STRUT
EA	8.66 kN	TIE
CF	10.39 kN	STRUT
FE	3.46 kN	TIE
FG	3.46 kN	STRUT
GA	12.12 kN	TIE
DG	24.25 kN	STRUT

Internal forces in a beam

BENDING STRESSES

When a beam is loaded it will deflect as in *Fig 16*. The material near the top of the beam becomes shorter and is in compression. The material near the bottom of the beam becomes longer and is in tension. There is a layer AB for which there is no change of

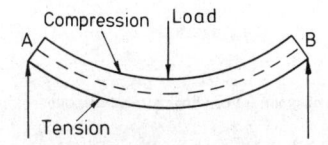

Fig 16 Deflection of a beam

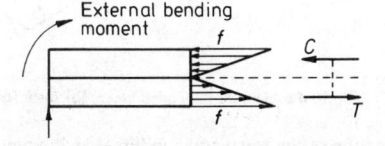

Fig 17 Internal forces due to external bending moment in a beam section

length which is thus unstressed. The maximum compressive and tensile stresses due to the bending occur at the top and bottom of the beam respectively.

Consider a section of the beam as shown in *Fig 17*. For the section of the beam to be in equilibrium the external bending moment must be balanced by the moment of the internal tensile and compressive forces. The compressive stress acting across the top of the beam section can be represented by a compressive force C. Similarly the tensile forces can be represented by the force T. These forces constitute a couple, the moment of which must balance the external bending moment.

In the design of beams it is essential that the compressive and tensile stresses induced by the bending do not exceed the permissible values. It may be shown that, for a rectangular beam section of breadth, b, and depth, d, that the maximum stress due to bending, f, is given by the formula:

$$M = f\frac{bd^2}{6}$$ where M is the external bending moment.

Naturally the maximum stress will be occasioned at that section of the beam where the external bending moment is a maximum.

> **Problem 20.** A timber beam 75 mm in breadth and 150 mm in depth spans 3 m. The beam carries a concentrated load of 4 kN at its centre. Calculate the maximum bending stress induced by this load.

The shear force and bending moment diagrams are illustrated in *Fig 18*. The maximum bending moment occurs at midspan and has a value of 3 kNm. As the dimensions of the cross

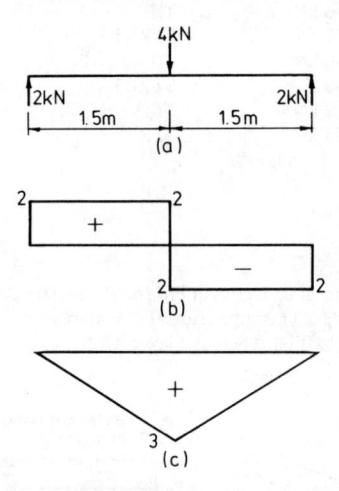

Fig 18 Problem 20 (a) loaded beam; (b) shear force diagram; (c) bending moment diagram

section of the beam are in millimetres it would be advisable to convert the bending moment in Nmm.

$$3 \text{ kNm} = 3000 \text{ Nm} = 3\,000\,000 \text{ Nmm}$$

For the given section:

$$\frac{bd^2}{6} = \frac{75 \times 150^2}{6} = 281\,250 \text{ mm}^3$$

Thus using the formula $M = f\dfrac{bd^2}{6}$

$$f = \frac{M}{bd^2/6} = \frac{3\,000\,000}{281\,250} = 10.67 \text{ n/mm}^2$$

The maximum bending stress is **10.67 N/mm²**.

Problem 21. A rectangular timber beam 75 mm in breadth is to carry a centre point load of 4 kN on a span of 3 m. Determine the required depth of the beam if the maximum bending stress is to be 7 N/mm².

The load situation is the same as in the last example and is illustrated in *Fig 18*. As before the maximum bending moment is 3 000 000 Nmm.

Applying $M = f \dfrac{bd^2}{6}$ $3\,000\,000 = 7 \times \dfrac{75 \times d^2}{6}$

since the maximum f is to be 7.   Hence $d^2 = \dfrac{6 \times 3\,000\,000}{7 \times 75} = 34285$ mm²

The depth of the beam is thus: $d = \sqrt{34285} = 185.2$ mm

SHEAR STRESSES

The shear force due to external loads causes a tendency for a beam to fail as shown in *Fig 19*. Thus vertical shear stresses will exist at any cross-section of the beam. Vertical shear stresses are always accompanied by horizontal shear stresses. The existence of this tendency to horizontal shear can be illustrated by bending a beam made up of a num-

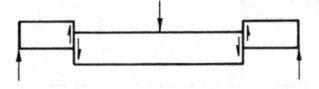

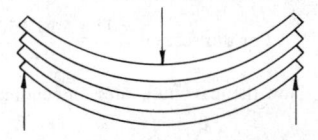

Fig 19 (above left) Vertical shear induced by external loads

Fig 20 (above right) Horizontal shear induced by external loads

Fig 21 (right) Horizontal and vertical shear forces acting on an element

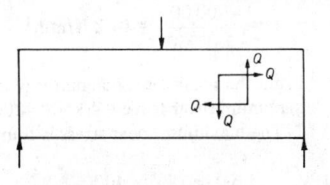

ber of layers; a pack of cards will demonstrate this adequately. The result of such bending is shown in *Fig 20*.

Fig 21 illustrates the horizontal and vertical shear forces acting on a small cube of material within the beam. The horizontal and vertical shear forces must have the same value otherwise the cube will rotate.

For a rectangular section the maximum value of the shear stress is given by

$$\tau = \dfrac{1.5 Q}{bd}$$

where τ is the shear stress, Q the shear force, d the depth of the beam and b its breadth.

The shear stresses create diagonal compression and tension within the material. If the shear forces acting on the cube in *Fig 21* are combined using the parallelogram of forces the result is shown in *Fig 22(a)*. The resulting forces tend to deform the cube as

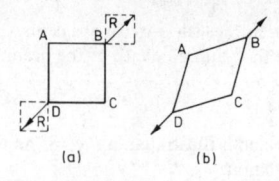

Fig 22 (a) Resultant of shear forces; (b) deformation due to shear forces

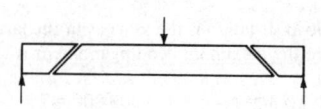

Fig 23 (below) Cracking due to induced tension

shown in *Fig 22(b)*. It will be seen that the material is in tension in direction BD and compression in the direction AC. If the material is weak in tension then it may fail by cracking at right angles to the tension force as shown in *Fig 23*. Thus shear reinforcement may be provided in reinforced concrete beams.

> *Problem 22.* A timber beam 75 mm in breadth and 150 mm deep carries a centre point load of 8 kN on a span of 2 m. Determine the maximum bending stress and the maximum shear stress.

Fig 24 shows the bending moment diagram from which it is observed that the maximum bending moment is 4 kNm = 4 000 000 Nmm.

For this section $\dfrac{bd^2}{6} = \dfrac{75 \times 150^2}{6} = 281250$

Now the maximum stress in bending is given by

$$f = \dfrac{M}{bd^2/6}$$

$$= \dfrac{4\,000\,000}{281\,250} = 14.2 \text{ N/mm}^2$$

From the shear force diagram it is seen that the maximum shear force = 4 kN = 4000 N.

The maximum shear stress is found using

$$\tau = \dfrac{1.5Q}{bd} = \dfrac{1.5 \times 4000}{75 \times 150} \text{ N/mm}^2 = 0.53 \text{ N/mm}^2$$

This example illustrates that in the majority of cases for steel and timber beams the stresses induced by bending are more important than the shear stresses.

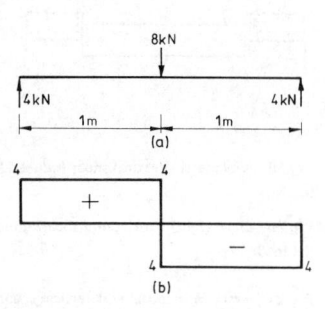

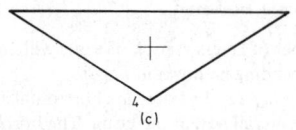

Fig 24 Problem 22 (a) loaded beam; (b) shear force diagram; (c) bending moment diagram

Internal forces in a column

If a short column is loaded axially it will shorten as predicted earlier in this chapter, at the same time it will increase in cross-sectional area as shown in *Fig 25*. The stress on any plane at right angles to the axis of the column is purely compressive. Shear stresses

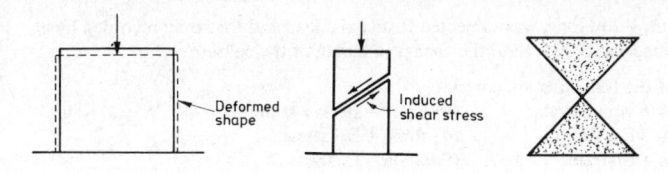

Fig 25 Deformed shape of a loaded short column

Fig 26 (above centre) Induced shear stress in a loaded short column

Fig 27 (above right) Failure of a concrete cube in compression

Fig 28 (right) Buckling of a slender column

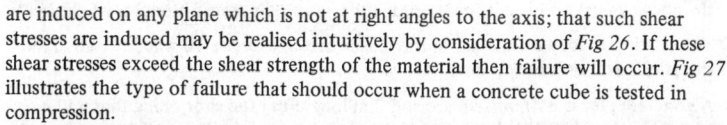

are induced on any plane which is not at right angles to the axis; that such shear stresses are induced may be realised intuitively by consideration of *Fig 26*. If these shear stresses exceed the shear strength of the material then failure will occur. *Fig 27* illustrates the type of failure that should occur when a concrete cube is tested in compression.

For long slender columns the axial load that can be sustained is limited by buckling. The critical buckling load for a column is beyond the scope of this book. The buckling of a column is shown in *Fig 28*.

Exercises (*Answers on page 106*)

1. A rectangular section, 100 mm × 150 mm, is subjected to an axial compressive load of 300 kN. Calculate the normal compressive stress in N/mm².

2. Which of the following are correct:
   (a) stress = area × force;
   (b) force = area × stress;
   (c) stress = force/area;
   (d) area = force/stress;
   (e) stress = area/force;
   (f) force = stress/area.

3. Calculate the tensile force that would create a normal tensile stress of 80 N/mm² in a 20 mm square tie bar.

4. A concrete cube of side 100 mm was tested in compression and failed at a load of 450 kN. Determine the compressive strength of the concrete in N/mm².

5. A rivet in single shear is to transmit a force of 40 kN. If the shear stress is not to exceed 100 N/mm² determine the minimum cross-sectional area of the rivet.

6. Which of the following are correct:
   (a) strain = original length × change in length
   (b) strain = original length/change in length
   (c) original length = strain/change in length
   (d) change in length = strain × original length
   (e) change in length = strain/original length.

7. A steel rod of length 200 mm was subjected to an axial tensile force and extended 0.25 mm. Calculate the strain.

8 A column, 1.2 m long, was subjected to an axial load and the strain recorded by a strain gauge was 0.001. Find the change in length of the column.

9 Which of the following are correct:
   (a) E = stress/strain;
   (b) E = stress × strain;
   (c) stress = E × strain;
   (d) strain = E × stress;
   (e) stress = E/strain;
   (f) strain = E/stress.

10 In a tensile test on a 10 mm diameter steel bar the following values of load and extension were recorded.

load (kN)	4	8	12	16	20
extension (mm)	0.0121	0.0245	0.0363	0.0482	0.0606

The extension was measured on an original gauge length of 50 mm. Plot a graph of the above values and use it to determine the value of Young's modulus of elasticity.

11 The maximum load on the bar in the tensile test in exercise 10 was 31.5 kN. Find the tensile strength of the steel.

12 A concrete pier is 800 mm square and 2 m long. Find the shortening that will occur when a load of 1500 kN is applied if the modulus of elasticity is 18 kN/mm^2.

13 Which of the following are correct:
   (a) permissible stress = ultimate stress × factor of safety
   (b) permissible stress = ultimate stress/factor of safety
   (c) factor of safety = ultimate stress × permissible stress
   (d) ultimate stress = factor of safety × permissible stress.

14 Find the maximum load that can be carried by a 25 mm square steel tie if the tensile strength of the steel is 450 N/mm^2 and a factor of safety of 2 is adopted.

15 The permissible stress in a concrete member is to be 9 N/mm^2.
   The compressive strength of the concrete is 28 N/mm^2.
   Find the factor of safety.

16 A beam, simply supported at its ends, spans 6 m and carries a point load of 5 kN at a distance of 4 m from the left hand end. Calculate the reactions and draw the shear force and bending moment diagrams.

17 A beam, simply supported at its ends, spans 8 m and carries point loads of 5 kN and 8 kN at distances of 2 m and 5 m from the left-hand end respectively. Calculate the reactions and draw the shear force and bending moment diagrams.

18 Select the correct option: the maximum bending moment always occurs:
   (a) where the shear force is a maximum;
   (b) where the shear force is zero;
   (c) at the right hand support;
   (d) at mid span.

19 Select the correct option: for a simply supported beam the maximum shear force always occurs:
   (a) at mid span;
   (b) near a support;
   (c) where the bending moment is a maximum.

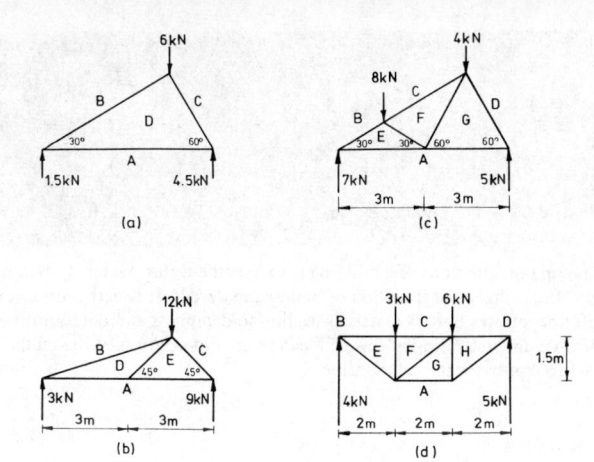

Fig 29 Diagrams for exercise 21

20 A simply supported beam, having a span of 4 m carries a centre point load. If the maximum bending moment is 16 kNm find the value of the centre point load.

21 Find the magnitude and type of force acting in each member of the frames illustrated in *Figs 29(a), (b), (c) and (d)*.

22 A rectangular timber beam spans 3 m and carries point loads of 1 kN and 2.5 kN at distances of 1 m and 2 m from the left hand end respectively. Calculate the reactions and find the maximum shear force and bending moment. If the beam is 100 mm in breadth and the maximum stress in bending is not to exceed 9 N/mm² find the necessary depth of the beam. Calculate also the maximum shear stress in the timber.

23 A timber beam, having a breadth of 50 mm and depth of 150 mm, is simply supported over a span of 2 m. Find the maximum centre point load if the stress in bending is not to exceed 7 N/mm².

4 Water and building

One of the major functions of a building is to exclude water. Water is often a most important factor in the degradation of building materials. It is necessary to consider how water penetrates porous materials leading to dampness and deterioration of materials. An understanding of this will lead to an appreciation of the methods employed to combat water penetration.

Surface tension

Attractive forces exist between molecules, if the molecules are of the same type they are termed cohesive forces. If the molecules are of different types the attractive forces are termed adhesive forces.

At surfaces between materials these forces give rise to a surface tension. *Fig 1* shows the surface between a liquid and a vapour. Below the surface a molecule such as X is attracted by all the molecules around it and the resultant force is zero. For the molecule Y in the surface the attraction forces to other molecules in the liquid are much greater than those to the molecules in the vapour. This arises since the molecules in the vapour are widely dispersed. There exists a net downward force trying to pull the molecule Y into the liquid. This is the case for all molecules in the surface. The molecules in the surface are not pulled into the bulk of the liquid since liquids are not readily compressed, however these forces do attempt to reduce the surface area of the liquid to a minimum and the surface is under tension and acts like a skin. That a surface tension exists and endeavours to reduce the surface area to a minimum is evidenced by a number of everyday occurrences.

The pond skater is supported on the water surface by the surface tension forces; a clean needle can be supported on the surface of water. Raindrops and soap bubbles endeavour to minimise their surface area by becoming as near spherical as gravity will permit. To increase the surface area of a liquid more molecules must be pushed into the surface against the attractive forces between the molecules and energy is required

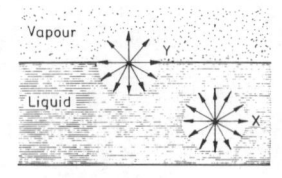

Fig 1 Surface tension due to molecular attraction

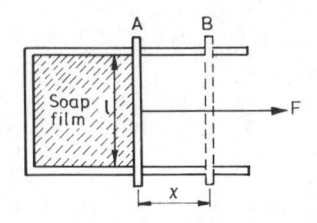

Fig 2 Surface energy of a soap film

to do this. The energy required to produce one square metre of a surface, at constant temperature, is called the surface energy, G.

The concept of surface energy is readily understood by considering the soap film and simple frame shown in *Fig 2*. When the force F pulls the slider from A to B then:

Work done = force × distance = Fx

Since the soap film has two sides the area of surface produced = $2lx$. Thus:

$$\text{Surface energy } G = \frac{\text{work done}}{\text{area of surface produced}}$$
$$= \frac{Fx}{2lx} = \frac{F}{2l}$$

F is the force which will keep the soap film extended at the position B and is balanced by the surface tension forces acting within the liquid surface and parallel to it.

The value of the force per unit length is termed the surface tension γ: $\gamma = F/2l$. Thus it appears that the surface tension and the surface energy are the same, this is not strictly true since temperature changes in extending the film have been ignored. In this simple discussion the surface energy and surface tension will be treated as the same. Surface tensions exist at surfaces between solids and gases and solids and liquids.

A method of measuring the surface tension of a liquid is to blow a bubble under the surface of a liquid. The stages in the development of the bubble are shown in *Fig 3*. It can be shown that the excess pressure required to produce a curved surface of radius, R, is given by:

$$p = \frac{2\gamma}{R}$$

where p is the excess pressure and γ is the surface tension of the liquid-vapour interface. As can be seen from the diagram the radius of curvature of the bubble is a minimum when the bubble is hemispherical and its radius is equal to the radius of the tube. At this stage the excess pressure is at a maximum.

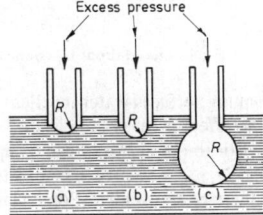

Fig 3 Development of a bubble

The pressure needed to maintain the bubble in *Fig 3(c)* is less than the maximum. If the pressure is maintained at the maximum value the bubble expands rapidly and breaks away. It is important to realise that a liquid surface can only be maintained in a curved state if an excess pressure exists on the inside of the curved surface.

Problem 1. An excess pressure of 290 N/m² was required to blow a bubble just below the surface of water using a 1 mm diameter glass tube. Calculate the surface tension between the water-air interface.

$$p = \frac{2\gamma}{R}$$

Transposing this formula:

$$\gamma = \frac{pR}{2} = \frac{290 \times 0.0005}{2} = 0.0725 \text{ N/m}$$

Contact angle

If a liquid drop is placed on a solid surface it can remain in equilibrium in either of the positions illustrated in *Fig 4*. There are three surface tension forces to consider:

γ_{sv} between the solid and vapour;
γ_{sl} between the solid and liquid;
γ_{lv} between the liquid and vapour; this is the surface tension considered in the last section.

If the drop remains in equilibrium then

$$\gamma_{sv} = \gamma_{sl} + \gamma_{lv} \cos \theta$$

θ is called the angle of contact. If θ is less than $90°$ the liquid is said to wet the surface. If θ is greater than $90°$ the liquid does not wet the surface. Silicone materials used for

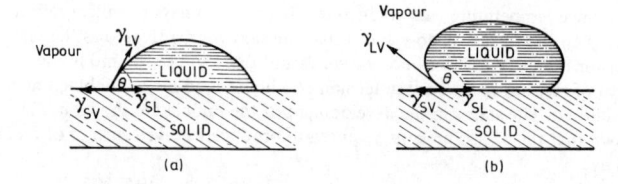

Fig 4 Liquid drops (a) contact angle less than $90°$; (b) contact angle greater than $90°$

making surfaces water repellent work by creating an angle of contact greater than $90°$ The effect is easily seen in the behaviour of rain drops on a newly polished car.

For the liquid to spread completely acrosss the surface θ must become zero hence if $\gamma_{SV} > \gamma_{SL} + \gamma_{LV}$ then complete spreading occurs. This is important for the spreading of adhesives and solders. Some plastics cannot easily be glued since their surface tension is so low that the adhesive will not spread.

Capillarity

For water in contact with many solids the angle of contact is less than $90°$ so that the water rises at a vertical boundary to form a **meniscus** as shown in *Fig 5(a)*. If a narrow tube is used the water is seen to rise up the tube as in *Fig 5(b)*; this is termed capillary action.

It is useful to evaluate the height to which water will rise in a given tube. Rather than attempt to evaluate all the forces involved it is easier to consider the pressure excess that causes the surface of the water in the tube to be curved. For simplicity consider the surface of the water in the tube shown in *Fig 6(a)* to be a hemisphere, that implies that the angle of contact is zero.

Considering *Fig 6(a)*. Since the water surface is curved there is a greater pressure at B, just above the surface, than there is at A, just below the surface. The pressure at B exceeds that at A by $2\gamma/R$, where R is the radius of the tube. The pressures at B and C are both the same being at atmospheric pressure. Hence the pressure at A must be less than atmospheric by $2\gamma/R$. Thus A and C cannot remain at the same level. The

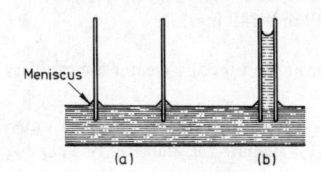

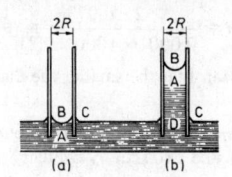

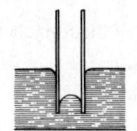

Fig 5 (above) (a) Formation of a meniscus; (b) capillary rise

Fig 6 (above right) Evaluation of capillary rise

Fig 7 (below right) Capillary depression

water column must rise, see *Fig 6(b)*, until the pressure at D is the same as that at C. Denote the atmospheric pressure by P, then:

Pressure at D = pressure at A + water pressure of column A to D

$$= P - \frac{2\gamma}{R} + h\rho g, \text{ where } \rho = \text{density of water.}$$

Pressure at C = P

The pressures at C and D are equal when

$$P = P - \frac{2\gamma}{R} + h\rho g$$

$$\frac{2\gamma}{R} = h\rho g$$

This, on rearrangement, gives

$$h = \frac{2\gamma}{R\rho g}$$

It may be shown that, if the angle of contact is θ then the above expression becomes:

$$h = \frac{2\gamma}{R\rho g} \cos\theta$$

If θ is greater than 90°, and the water does not wet the surface, then $\cos\theta$ becomes negative and capillary depression occurs as shown in *Fig 7*. A number of materials, including silicones are used for preventing capillary action.

> **Problem 2.** Find the capillary rise of water in a glass tube of (a) 1.0 mm (b) 0.2 mm diameter. Assume the surface tension of water to be 0.073 N/m and the angle of contact to be zero.

(a) Using $h = \dfrac{2\gamma}{R\rho g}$

$$h = \frac{2 \times 0.073}{0.0005 \times 1000 \times 9.81} = 0.03 \text{ m} = \mathbf{30 \text{ mm}}$$

(b) $$h = \frac{2 \times 0.073}{0.0001 \times 1000 \times 9.81} = 0.149 \text{ m} = \textbf{149 mm}$$

It will be seen that the smaller the diameter of the tube the greater the capillary rise.

> *Problem 3.* A brick stands on end in a shallow tray of water and the water is seen to rise 180 mm by capillary action. Calculate the diameter of a capillary that would account for this rise assuming the surface tension of water is 0.073 N/m and that the angle of contact is 10°

Transposition of the formula for capillary rise gives:

$$R = \frac{2\gamma \cos \theta}{h\rho g}$$

$$R = \frac{2 \times 0.073 \times \cos 10°}{0.180 \times 1000 \times 9.81}$$

$$= 0.000\,081 \text{ m} = 0.081 \text{ mm}$$

Thus diameter = $2 \times 0.081 = \textbf{0.162 mm}$

Practical measures for preventing moisture penetration

Whilst studying the Construction Technology unit the reader will have encountered many methods by which capillary action is prevented. The following are a few of the more obvious methods.
(a) the use of impervious materials for damp proof courses and damp-proof membranes. Such materials include polythene, lead, bitumen products and slate.
(b) the use of silicones and other water repellent materials for injecting damp-proof courses and for making renderings water repellent.
(c) The grooves cut in the underside of window sills and copings to prevent water spreading along the underside and thus into contact with the face of the building.
(d) anti-capillary grooves in window frame construction.
(e) a twist or similar device in the centre of a wall tie to prevent water spreading across to the inner leaf.
(f) the curvature of clay roofing tiles to prevent capillary rise between them.

Occasionally capillary action is useful as in capillary plumbing fittings where capillary action is used to spread the solder completely between the fitting and the pipe thus ensuring that the joint is watertight. It is essential that all surfaces are thoroughly clean and free of grease to ensure spreading of the solder.

Porosity of building materials

Voids or pores in materials have a number of effects on their properties. Voids reduce the density and the strength of a material. For example one per cent voids in concrete will reduce the strength by approximately 5%. If the pores are interconnected then capillary action can take place. Materials of a cellular nature, for example some foamed plastics, in which the cells are not interconnected will not allow the passage of water. The amount of water that a material will absorb will depend on the volume of the interconnected pore space.

The porosity of a material will influence its weathering characteristics and its resistance to freezing. Water expands by about 10% on freezing and large forces are developed if this expansion is restrained. Thus if the water in the pore space of a material is frozen sufficient pressure may be developed to disrupt the material. The exact nature of the relationship between frost resistance and the volume of pores and their size distribution is complicated. The volume of pore space by itself is not an accurate measure of frost resistance. In the case of concrete for roads frost resistance is increased when a small percentage of air is entrained. Since air has a very low thermal conductivity porous materials make good thermal insulators.

In discussing the density of porous materials care must be exercised to avoid confusion. The term bulk density is used to describe the density of the material including the air voids:

$$\text{bulk density} = \frac{\text{mass}}{\text{total volume}}$$

The term solid density describes the density of the solid part of the material:

$$\text{solid density} = \frac{\text{mass}}{\text{volume of solids}}$$

In some branches of materials science the relative density of the solids is used:

$$\text{relative density} = \frac{\text{density of solids}}{\text{density of water}}$$

Thus it will be seen that: solid density = density of water × relative density.

The porosity of a material is defined as:

$$\text{porosity} = \frac{\text{volume of voids}}{\text{total volume}} = 1 - \frac{\text{bulk density}}{\text{solid density}}$$

Problem 4. The volume of a piece of brick was found to be 240 ml. When this piece of brick was crushed the volume of the solids was found to be 180 ml. Calculate the porosity.

Volume of voids = 240 − 180 = 60 ml

$$\text{Porosity} = \frac{\text{volume of voids}}{\text{total volume}} = \frac{60}{240} = 0.25 \text{ or } 25\%$$

Problem 5. It was found that 8.625 kg of aggregate filled a 5 litre container. Assuming that the relative density of the aggregate particles is 2.65 determine the bulk density of the aggregate and the porosity.

$$\text{Bulk density} = \frac{\text{mass}}{\text{total volume}} = \frac{8.625}{0.005} = 1725 \text{ kg/m}^3$$

Solid density = density of water × relative density
= 1000 × 2.65 = 2650 kg/m³

$$\text{Porosity} = 1 - \frac{\text{bulk density}}{\text{solid density}} = 1 - \frac{1725}{2650} = 0.35$$

Problem 6. A cube of a building stone of side 0.1 m had a mass of 1.82 kg when dry. When the void space was completely saturated by water the mass of the stone was 2.12 kg. Determine the bulk density when dry, the porosity and the solid density of the stone.

Bulk density = $\dfrac{\text{mass}}{\text{total volume}} = \dfrac{1.82}{0.1 \times 0.1 \times 0.1} = 1820 \text{ kg/m}^3$

Mass of water in void space = $2.12 - 1.82 = 0.3$ kg

Volume of water in void space = volume of voids

$$= \dfrac{\text{mass of water}}{\text{density of water}}$$

$$= \dfrac{0.3}{1000} = 0.0003 \text{ m}^3$$

Porosity = $\dfrac{\text{volume of voids}}{\text{total volume}} = \dfrac{0.0003}{0.1 \times 0.1 \times 0.1} = \dfrac{0.0003}{0.001} = 0.3$

Volume of solids = total volume − volume of voids
 = $0.001 - 0.0003 = 0.0007$ m^3

Solid density = $\dfrac{\text{mass}}{\text{volume of solids}}$

$= \dfrac{1.82}{0.0007} = 2600$ kg/m^3

Electrolytic corrosion

The corrosion of metals can occur by direct oxidation in a dry atmosphere or by electrolytic corrosion in a wet situation. In construction the second mechanism is the most important. In electrolytic corrosion there is a flow of electric current, electrons, from one part of a metallic system to another.

The following definitions are important:

Electrolyte: A fluid capable of conducting electricity, in building this is usually water containing dissolved salts.

Anode: That part of the metallic system from which electrons enter the external circuit and from which metallic ions, which are positively charged, enter into solution in the electrolyte. In a corrosion cell it is the anode which corrodes.

Cathode: That part of the metallic system by which the electrons leave the external circuit and return to the electrolyte. The cathode is protected in a corrosion cell.

ELECTRODE POTENTIALS

If a metal is placed in an electrolyte positive metal ions will dissolve in the electrolyte leaving electrons on the metal. For instance for zinc the reaction can be shown as:

$Zn \rightarrow Zn^{++} + 2e^-$

As more metal ions enter the solution the metal becomes negatively charged due to the remaining electrons. This negative charge opposes the release of further ions. The negative voltage developed, measured under standard conditions, is the electrode potential. In measuring the electrode potential hydrogen is used as the reference. The following table lists some electrode potentials:

Metal	electrode potential (V)
copper	+0.34
hydrogen	0
lead	−0.13
iron	−0.44
zinc	−0.76
aluminium	−1.67

If the electrons are not removed from the metal the solution process ceases, however if they are removed the metal will continue to dissolve. It is now necessary to consider the mechanisms by which electrons can be removed. If the electrolyte contains hydrogen ions, H^+, the metals below hydrogen in the above table can donate their electrons to the hydrogen ions and produce hydrogen gas:

$$2(H^+) + 2e^- = H_2$$

Acids contain a large number of hydrogen ions and the metal dissolves rapidly. Pure water does contain some H^+ ions but the corrosion rate is slow for most metals.

DISSIMILAR METALS

When two dissimilar metals in electrical contact are placed in an electrolyte the anodic metal will suffer more corrosion and the cathodic less than if they were placed unconnected in the same solution. The anodic metal will be the one with the greatest negative electrode potential, that is lower down the above table. *Fig 8* illustrates a corrosion cell; the iron will be the anode and the copper will be the cathode. At the anode the iron will go into solution yielding electrons:

$$Fe \rightarrow Fe^{++} + 2e^-$$

The electrons will flow to the cathode thus giving an electric current. This current will only continue to flow if oxygen is present when a reaction such as the following can occur:

$$O + H_2O + 2e^- = 2(OH)^-$$

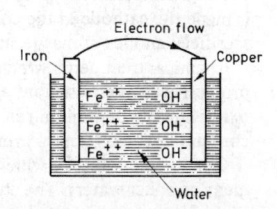

Fig 8   Iron–copper corrosion cell

The Fe^{++} ions will combine with the OH^- ions to give ferrous hydroxide, $Fe(OH)_2$ which will oxidise to yellow rust, $Fe_2O_3.H_2O$. If the oxygen supply is limited the end product can be black magnetite, Fe_3O_4, which is often found in central heating systems. Thus it can be seen that the presence of oxygen is very important in accelerating the corrosion process of many metals, particularly iron.

It will be appreciated that alloys, consisting of more than one metal, are likely to have less corrosion resistance than pure metals, the stainless steels are an important exception. It should also be noted that the table of electrode potentials is obtained under standard conditions and that the order of the metals can change in different situations.

OTHER CAUSES OF ELECTROLYTIC CORROSION

It is possible for anodes and cathodes to form on a single piece of metal and for corrosion to proceed in a damp environment. Differences in oxygen concentration will lead to an anode being formed at the oxygen deficient part. Thus corrosion can occur in crevices into which moisture can penetrate but in which the oxygen supply is limited. If variations occur in the concentration of the electrolyte an anode is formed in the region of low concentration. Anodes will be formed at stressed regions in the metal such as bends, rivet holes and grain boundaries.

At breaks in an oxide film on the metal an anode can be formed, the oxide film becoming the cathode. The rate of corrosion will be high since the cathode will have a large area compared with the anode. A large cathode area will always increase the rate of corrosion.

In waterlogged airless soils, which are usually non-corrosive the action of certain bacteria, which are capable of reducing sulphates, enable sulphates in the soil to take part in the corrosion process and for corrosion to proceed rapidly.

CORROSION PROTECTION

Paints and varnishes are often used to prevent corrosion. The aim is to produce an impervious film and several coats are usually necessary but even after this treatment the paint film will not be completely impervious. Additional protection can be achieved by a suitable choice of paint pigments such as the oxides of iron, zinc and titanium. Non porous films can be achieved with bitumen based paints.

The metal can be protected by a coating of a more anodic metal. In event of damage to the coating exposing the metal, the coating will corrode in preference to the metal. This is an example of cathodic protection and is typified by zinc and chromium coatings on steel. The phrase cathodic protection implies that the metal to be protected is made the cathode in the corrosion cell. Cathodic protection of steel by the use of sacrificial anodes of magnesium or zinc has been widely used.

Chemical treatments which produce a phosphate or chromate coating are reasonably protective on their own but are often used prior to painting. The naturally protective oxide film on aluminium can be thickened by anodising. Stainless steels incorporate chromium which forms a protective film of chromium oxide.

Corrosion inhibitors, which reduce the rate of corrosion are widely used in the treatment of boiler waters. The study of inhibitors is beyond the scope of this book but it should be noted that alkaline environments inhibit the corrosion of steel. The hydration products of Portland cement are alkaline and thus inhibit the corrosion of reinforcing steel.

Exercises (*Answers on page 108*)

1. An excess pressure of 145 N/m^2 was required to blow a bubble just below the surface of water using a tube of diameter 2 mm. When soap solution was added to the water the excess pressure required to blow a bubble was 45 N/m^2. Calculate the surface tension of water and soap solution.

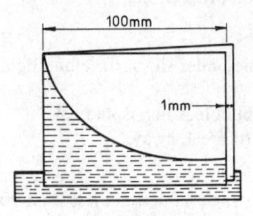

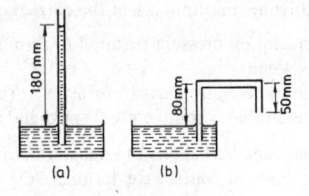

Fig 9 Exercise 2   **Fig 10 Exercise 3**

2 Two glass plates are in contact along one edge. The other edges are separated by a distance of 1 mm. This arrangement is placed in a tray of water as shown in *Fig 9*. The water rises by capillary action as shown. Calculate the capillary rise at intervals of 10 mm along the plate and illustrate the results graphically. Take the surface tension of water to be 0.073 N/m and the angle of contact as 0°.

3 A fine tube placed in water gave a capillary rise of 180 mm as shown in *Fig 10(a)*. If the tube is bent as shown in *Fig 10(b)* state whether water will flow from the capillary tube and thus empty the container. Explain the answer very carefully.

4 Draw labelled sketches of six methods by which capillary action is prevented in construction.

5 Calculate the pore diameter that would be necessary to cause water to rise 200 mm in brickwork if the surface tension of water is 0.073 N/m and the angle of contact is 5°. Explain the effect of treating the brickwork with a water repellent which increases the angle of contact to 105°.

6 A soil has a bulk density of 1760 kg/m^3 and a relative density of 2.65. Determine the porosity.

7 A metal container of volume 0.01 m^3 when filled with dry sand was found to contain 18 kg of sand. The container was stood on a level surface and water was poured in until the container was just on the point of overflowing. The volume of water added was 3.2 litres. Calculate the bulk density of the dry sand, the porosity and the solid density of the sand particles.

8 List six causes of electrolytic corrosion.

9 Explain each of the following effects:
   (a) Two steel plates are riveted together and used in a damp situation; severe corrosion occurred between the plates.
   (b) If oxide films produced during the manufacture of copper hot water cylinders are not removed there is a risk that small holes will occur in the cylinder due to corrosion.
   (c) A decorative plastic strip is fixed to the side of a car by means of metal clips through holes in the side of the car. After some time corrosion occurred at the fixing points.
   (d) The corrosion rate of a certain aluminium alloy in a marine environment was found to be much greater than that of pure aluminium.
   (e) The life of a steel boiler used in an indirect heating system is usually much greater than that used in a direct system.

10 Write a short account of the methods of corrosion protection.

In the following questions select the correct options.

11 The maximum pressure required to form a bubble under the surface of a liquid occurs when:
 (a) the bubble just starts to form; (b) the bubble is hemispherical;
 (c) the bubble is completely formed and about to break away.

12 A water repellent applied to a material causes:
 (a) an angle of contact greater than 90°; (b) an angle of contact less than 90°;
 (c) a decrease in the surface tension of the water-air interface; (d) an increase in the surface tension of the water-air interface.

13 Porosity in a building material:
 (a) decreases the strength; (b) decreases the thermal insulation;
 (c) increases the thermal insulation; (d) always decreases the frost resistance;
 (e) always creates capillary action with water.

14 The angle of contact of water and a plastic is 105°. If a small tube of the plastic is placed in water; the water in the tube:
 (a) rises; (b) depresses;
 (c) stays at the same level as in the container.

15 An anode in a corrosion cell is:
 (a) the metal that is protected; (b) the metal that corrodes;
 (c) the liquid conducting electricity; (d) the metal that enters solution.

16 Tin has an electrode potential of −0.14 V. If a break in the tin plate on a tin plated iron container occurs then:
 (a) the tin is the anode; (b) the tin is the cathode:
 (c) the tin corrodes; (d) the iron corrodes.

17 Iron alloys:
 (a) always corrode more rapidly than pure iron; (b) some times corrode more rapidly than pure iron;
 (c) always corrode less rapidly than pure iron; (d) sometimes corrode less rapidly than pure iron.

18 In installing a hot water system some small pieces of copper were accidentally left in the bottom of a zinc galvanised steel tank. The result would be:
 (a) the copper would corrode; (b) the zinc coating on the tank would corrode;
 (c) nothing will happen since zinc and copper resist corrosion.

19 If zinc covered iron nails were used to fix copper roof sheeting the expected result would be:
 (a) the copper sheet would corrode around the nails;
 (b) the nails would corrode rapidly; (c) the nails would be protected from corrosion;
 (d) nothing will happen since zinc and copper resist corrosion.

20 Corrosion proceeds most rapidly if there exists:
 (a) a large anode and a small cathode; (b) a small anode and a large cathode;
 (c) an absence of oxygen.

5 Cement, aggregates and concretes

Cements

MANUFACTURE OF PORTLAND CEMENT

Portland cement is made from chalk or limestone and clay or shale. The chalk or limestone is mainly calcium carbonate. The clay or shale contains silicates, aluminates and iron compounds, including iron oxide. The raw materials are finely broken up, intimately mixed and burnt in a rotary kiln at temperature of about $1400°$ C. The raw materials react chemically, partially fuse and emerge from the kiln as pellets of cement clinker. The clinker is then ground to a fine powder. At this stage a small percentage of gypsum is added in order to prevent the cement flash setting when mixed with water.

CHEMICAL COMPOSITION OF PORTLAND CEMENT

The chemical reactions in the kiln cause the calcium carbonate, silica, alumina and iron oxide to combine to give four major compounds:

    Tricalcium silicate $3CaO.SiO_2$
    Dicalcium silicate $2CaO.SiO_2$
    Tricalcium aluminate $3CaO.Al_2O_3$
    A ferrite phase which is approximately tetracalcium aluminoferrite:
                $4CaO.Al_2O_3.Fe_2O_3$

It is common in cement chemistry to use the following abbreviations:
    C for CaO (lime)
    S for SiO_2 (silica)   F for Fe_2O_3 (iron oxide)
    A for Al_2O_3 (alumina)   H for H_2O (water)

Thus the chemical formulae of the four major compounds can be written as:
    Tricalcium silicate C_3S Tricalcium aluminate C_3A
    Dicalcium silicate C_2S Ferrite phase C_4AF

HYDRATION OF PORTLAND CEMENT

When cement is mixed with water the compounds in the cement chemically react with water. For the tricalcium silicate and dicalcium silicate these reactions may be shown schematically by the following equations:

$$2C_3S + 6H \rightarrow C_3S_2H_3 + 3Ca(OH)_2$$
$$2C_2S + 4H \rightarrow C_3S_2H_3 + Ca(OH)_2$$

In each case the hydrated calcium silicates are the same. The calcium hydroxide, $Ca(OH)_2$, produced in each case ensures that the cement paste is alkaline which provides an environment in which reinforcing steel is protected from corrosion. Tricalcium

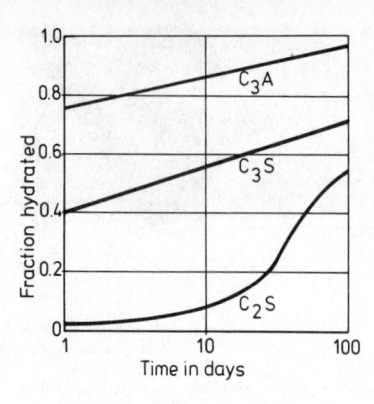

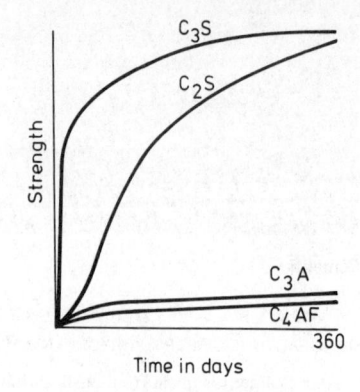

Fig 1 Rate of hydration of pure cement compounds

Fig 2 Rate of gain of strength of pure cement compounds

aluminate, C_3A, and water react very rapidly and cause an immediate setting of the cement paste. This rapid reaction is controlled by the addition of a small amount of gypsum. Gypsum is calcium sulphate dihydrate, $CaSO_4.2H_2O$ and reacts with the tricalcium aluminate to produce, initially, a calcium sulphoaluminate. The hydration reactions of the ferrite phase are of little importance. The ferrite phase being mainly responsible for the colour of the cement. If iron free clays are used, such as china clay, it is possible to make white Portland cement.

It is important to appreciate the rate of hydration and rate of gain of strength of the compounds in the cement; these are shown in *Figs 1 and 2*. From these graphs it will be seen that the strength achieved by the C_3A and C_4AF is small. The tricalcium silicate hydrates rapidly and thus gives rise to the early strength of the cement paste whereas the dicalcium silicate hydrates more slowly and thus contributes the later development of strength. Notice that the hydration reactions continue for a long period of time and it is essential to keep the concrete moist for as long as possible to ensure that it is properly cured and the strength allowed to develop. The rate of hydration decreases as the temperature decreases and ceases at the freezing point of water.

The hydration reactions of the cement compounds are exothermic, that is heat is produced during the hydration process. The heat of hydration of each of the cement compounds is:

Tricalcium silicate	502 kJ/kg	Tricalcium aluminate	865 kJ/kg
Dicalcium silicate	259 kJ/kg	Tetracalcium aluminoferrite	419 kJ/kg

TYPES OF CEMENT

By suitable modifications of the initial materials in the manufacturing process it is possible to produce cements with different properties to suit different conditions of use.

Ordinary Portland cement (BS 12:1978) Produces a medium rate of gain of strength, suitable for general concreting work.

Rapid hardening Portland cement (BS 12:1978) The initial setting time of this cement is the same as that of ordinary Portland cement but having set it gains strength more rapidly. The cement is more finely ground than ordinary Portland cement which allows the hydration process to proceed more rapidly since a greater surface area of the cement is exposed to water. In some instances it may contain a higher proportion of tricalcium silicate, C_3S, than is usually found in ordinary Portland cement.

Ultra high early strength Portland cement This is an extremely finely divided Portland cement which in the main complies with BS 12. The strength at 16 hours is twice that of rapid hardening Portland cement, but the strength at 28 days is similar. This cement is suitable for the production of precast and prestressed concrete and all situations where rapid striking of formwork is necessary.

Sulphate resisting Portland cement (BS 4027:1972) The hydrates of tricalcium aluminate in hardened concrete can be attacked by sulphates giving rise to a large increase in volume and a gradual disintegration of the concrete.

In the manufacture of sulphate resisting cement iron oxide is added so that the ferrite phase is formed in preference to tricalcium aluminate. Under BS 4027 the maximum C_3A content is limited to 3.5%. The rate of gain of strength is initially low since the cement contains a high proportion of dicalcium silicate, this is partially offset by finer grinding than for ordinary Portland cement. The cement would be used in situations where sulphate attack is likely, for instance sulphate bearing groundwater and marine structures.

Low heat Portland cement (BS 1370:1974) In mass concrete work the rise in temperature caused by the heat produced by the hydration of the cement can lead to serious cracking. As stated previously, the heats of hydration of tricalcium silicate and tricalcium aluminate are higher than the other compounds. Thus low heat Portland cement contains a much lower content of tricalcium silicate and tricalcium aluminate than ordinary Portland cement. As the content of dicalcium silicate is high the early strength development is slow, this is only partially offset by finer grinding. The final strength is similar to ordinary Portland cement. Since the tricalcium aluminate content is low the cement has good sulphate resistance.

White cement White cement is produced by using white china clay in place of the more readily obtainable brown clay. An oil fired, rather than a coal fired, kiln is used. Since this cement is expensive it is only used where specific architectural effects are required.

Masonry cement (BS 5224:1976) This cement is intended for making brick laying mortars and is manufactured by intergrinding ordinary Portland cement clinker, an inert filler and an air entraining agent.

Portland blastfurnace cement (BS 146:1973) This cement is made by grinding together Portland cement clinker, up to 65% quenched blast-furnace slag and gypsum. The blast furnace slag, which is a by-product of iron making, contains lime, silica and alumina, but not in the same proportions as Portland cement. The Portland cement component hydrates in the usual manner and it is probable that the calcium hydroxide thus produced initiates the hydration of the blast-furnace slag.

The rate of hardening of this cement is somewhat slower than ordinary Portland cement in the early stages but the final strengths are comparable. Adequate curing is thus essential. Since Portland blast-furnace cement has a fairly high sulphate resistance it is often used for marine construction.

Low heat Portland blast-furnace cement (BS 4246:1974) This is similar to Portland blast-furnace cement describe above. The heat of hydration and strength development are similar to low heat Portland cement.

Supersulphated cement (BS 4248:1974) This is not a Portland cement and is made by intergrinding not less than 75% granulated slag, calcium sulphate and about 5% Portland cement clinker. Supersulphated cement is highly resistant to sea water and to most concentrations of sulphates found in soils. It also has a good resistance to weak acids. It can thus be used for marine work, concrete in sulphate bearing soils and for sewer construction. The heat of hydration is low so that it is suitable for mass concrete work but care must be taken in cold weather since the rate of gain of strength is reduced considerably at low temperatures.

In all applications wet curing for at least four days is essential to prevent the surface of the concrete from powdering. The final strength achieved is similar to ordinary Portland cement. This cement is no longer made in Great Britain.

High alumina cement (BS 915:1972) This is not a Portland cement. It is made from limestone and bauxite. The limestone is nearly pure calcium carbonate and the bauxite is nearly pure alumina. The raw materials are melted in a furnace at $1600°$ C. The molten material is slowly cooled and then crushed and ground. The resulting cement consists largely of monocalcium aluminate, CA.

Although the initial setting time of high alumina cement is about 4 to 5 hours the rate of gain of strength after this is very rapid. 80% of the final strength being achieved in the first 24 hours. The final strength achieved for high alumina cement concrete is typically at least twice that of an ordinary Portland cement concrete. It is thus possible for the formwork to be struck after only 6 to 8 hours. Since the hydration process is very rapid the rate at which heat is evolved is also rapid so high alumina cement is only suitable for work that will be cast in thin sections or layers.

High alumina cement has a very good resistance to sulphate attack and a fair resistance to most very dilute acids. Caustic alkalis attack the cement vigorously. If the hardened concrete is subjected to temperatures in excess of $25°$ C and high humidity the hydration products convert to a different structure with a loss of strength. The remaining strength may be as little as 30 per cent of the original strength. The lower the amount of water in the original mix the higher the residual strength. The converted form of the material is also liable to sulphate attack. A number of building failures have, at least in part, been due to the use of high alumina cement concretes in warm, moist environments.

Mixtures of ordinary Portland cement and high alumina cement in suitable proportions will give a flash set. The strength of such mixtures is low but they may be employed in emergency work, for instance, on drains. In normal practice it is essential to ensure that these two cements are not mixed. No additives are allowable with high alumina cement.

> *Problem 1.* The table shows the composition of two cements. Cement A is known to be ordinary Portland cement. Deduce the properties of cement B and state which cement it is likely to be.

Compound (%)	Cement A	Cement B
tricalcium silicate	49	55
dicalcium silicate	25	16
tricalcium aluminate	11	11
ferrite phase	9	9
other materials	6	9

Cement B has a higher proportion of tricalcium silicate and less dicalcium silicate than cement A. It will thus develop strength much more rapidly and is likely to be rapid hardening Portland cement. For the same reason it will have a higher heat of hydration. Since the tricalcium aluminate content is the same in both cases the sulphate resistance will be similar.

> *Problem 2.* A cement C has the following composition by percentage:
>
> $C_3S = 43$; $C_2S = 37$; $C_3A = 2$; $C_4AF = 14$
>
> Deduce the type of cement by comparing it with cement A in *Problem 1*.

Since the C_3S content is less, the C_2S content more and the percentage of C_3A is less than 3.5 per cent it would appear that the cement is a sulphate resisting Portland cement.

Concrete aggregates

AGGREGATES FROM NATURAL SOURCES

Size and grading In a typical concrete the aggregate will occupy about 75% of the total volume. The properties of the aggregate are thus of the utmost importance. Much of the aggregate used is obtained from natural stones which may be crushed or uncrushed.

A coarse aggregate is one that is mainly retained on a 5.00 mm sieve. BS 882:1973 gives the permissible grading for graded coarse aggregates of nominal size 40 mm to 5 mm, 20 mm to 5 mm and 14 mm to 5 mm. Gradings are also specified for nominal single-sized aggregates of size 63 mm, 40 mm, 20 mm, 14 mm and 10 mm. As an example *Table 1* gives the percentage by weight passing the specified sieve for a 20 mm to 5 mm graded aggregate.

Table 1

Sieve size (mm)	37.5	20.0	10.0	5.00
% passing	100	95–100	30–60	0–10

Table 2 Percentage by weight passing BS sieves

BS 410 test sieve	Grading Zone 1	Grading Zone 2	Grading Zone 3	Grading Zone 4
mm				
10.0	100	100	100	100
5.00	90–100	90–100	90–100	95–100
2.36	60–95	75–100	85–100	95–100
1.18	30–70	55–90	75–100	90–100
μm				
600	15–34	35–59	60–79	80–100
300	5–20	8–30	12–40	15–50
150	0–10	0–10	0–10	0–15

It will be seen that all particles must be smaller than 37.5 mm, the majority, in excess of 95%, must pass the 20.0 mm sieve and that less than 10% can be smaller than 5.00 mm.

A fine aggregate is one that mainly passes a 5.00 mm sieve. Fine aggregates are classified into four zones depending upon their grading. The percentage by weight passing the various BS sieves is shown in *Table 2*, which is taken from BS 882.

The grading of a fine aggregate shall be within the limits of one of the zones except that a total tolerance of up to 5% may be applied to the percentages given in light type. This tolerance may be split up, for example it could be 1% on each of three sieves and 2% on another.

> *Problem 3.* In performing a sieve analysis on a fine aggregate 200 grammes of the material were taken. The mass of aggregate retained on each sieve was determined and is shown in the table. In addition it was found that 6 grammes passed the 150 μm sieve.
>
> | Sieve size (mm) | 10.0 | 5.00 | 2.36 | 1.18 | 600 μm | 300 μm | 150 μm |
> |---|---|---|---|---|---|---|---|
> | Mass retained (g) | 0 | 6 | 10 | 12 | 46 | 84 | 36 |

Determine the grading zone of the fine aggregate.

In performing a sieve analysis it is usual to check that the sum of the weight retained plus that passing the 150 μm sieve is equal to the original mass. This ensures that the masses are correct and that no material has been lost:

$6 + 10 + 12 + 46 + 84 + 36 + 6 = 200$

It is now necessary to convert from the mass of material retained on each sieve to the mass of material passing that sieve. As no material is retained on the 10.0 mm sieve then 200 g passes this sieve. The 5.00 mm sieve retains 6 g so that 194 g pass the sieve. Of this 194 g of sand 10 g is retained on the 2.36 mm sieve therefore 184 g passes this sieve. This process is continued for the rest of the sieves. The mass passing each sieve is then expressed as a percentage of the total mass of 200 g. The results of these calculations are as follows:

Sieve size	Mass retained (g)	Mass passing (g)	Percentage passing
mm			
10.0	0	200	100
5.00	6	194	97
2.36	10	184	92
1.18	12	172	86
μm			
600	46	126	63
300	84	42	21
150	36	6	3

From an inspection of *Table 2* for the grading zones it will be seen that the fine aggregate is in zone 3. It is instructive to show the results graphically as shown in *Fig 3*.

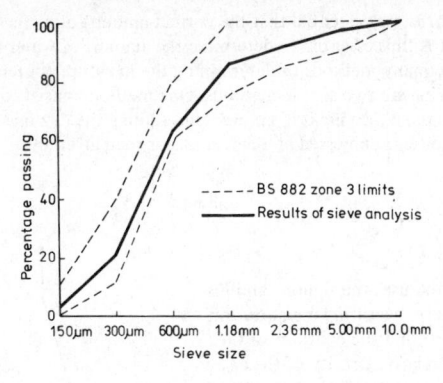

Fig 3 Grading of fine aggregate in example 3

An all-in aggregate is composed of a mixture of coarse and fine aggregates. BS 882 gives the permissible gradings for 40 mm and 20 mm all-in aggregates.

TYPES OF NATURAL AGGREGATE

Coarse aggregates may be described as uncrushed gravel, crushed gravel or crushed stone. Similarly fine aggregate may be described as natural sand, crushed gravel sand or crushed stone sand. Many types of gravel and rock make suitable concrete aggregates and the following are some of the more common group names: flint, limestone, granite, basalt, gritstone and gabbro.

Shape The shape is defined by one of the following classifications: rounded, irregular, angular, flaky, elongated and flaky and elongated. Rounded aggregates are typically river or seashore gravels, irregular aggregates are other gravels such as land dug flints and angular aggregates with well defined edges are typified by crushed rocks.

Surface texture The surface texture is classified as follows: glassy, smooth, granular, rough, crystalline or honeycombed.

Silt, clay and dust The presence of very fine material coating the surface of the aggregate particles will reduce the bond strength between the cement and the aggregate. Loose fine particles will require an increase in the water content of the mix because of their large surface area. Silt is a natural fine material with a particle size between 60 μm and 2 μm. Clay has a particle size less than 2 μm. Dust can arise from very fine particles of crushed rocks.

The field settling test is an approximate method for estimating silt, clay and dust content; the test is not suitable for crushed stone sands or coarse aggregates. 50 ml of 1% salt solution are poured into a 250 ml measuring cylinder. The sand is added until its volume is 100 ml. More salt solution is added until the final volume is 150 ml. The measuring cylinder is shaken vigorously, placed on a level surface, tapped until the sand surface is level and allowed to stand for three hours. The height of the silt layer is then expressed as a percentage of the height of the sand below the silt-sand interface. If this result exceeds 8 per cent then a more detailed test must be carried out to assess the suitability of the aggregate.

Moisture content It is very essential that the correct amount of water is added to a concrete mix and it is thus essential to determine the amount of water that is in the aggregate. There are many methods of determining the moisture content of an aggregate of which the following are two simple methods. One method consists of taking a sample of known mass, W, drying it in an oven and finding the dry mass of the material, W_s. The moisture content expressed as a percentage is then given by:

$$\text{moisture content} = \frac{W - W_s}{W_s} \times 100$$

The second method uses the siphon can illustrated in *Fig 4*. Before measuring moisture content it is necessary to find the constant of the siphon can and the relative density of the aggregate. The constant is measured by filling the can with water, running off the water through tap A and then measuring the volume of water discharged when tap B is opened. The volume of the water collected is the required constant. In order to determine the relative density it is necessary to find the volume of a known mass of dry aggregate. The water in the siphon can is allowed to fall to the level of tap B and the taps are closed; 2000 grammes of dry aggregates are poured in and stirred to expel air. The volume of water collected when tap A is opened is found. The total volume of the aggregate is this volume plus the constant of the siphon can.

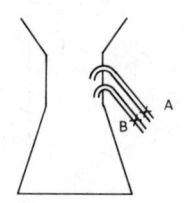

Fig 4 Siphon can

The relative density of the aggregate, G_s, is given by:

$$G_s = \frac{\text{mass of aggregate in grammes}}{\text{volume of aggregate in ml}}$$

To find the moisture content of a wet aggregate, the volume of 2000 grammes of the wet aggregate is determined as above. Then the moisture content, as a percentage, is given by:

$$\text{moisture content} = \frac{G_s V - W}{G_s (W - V)} \times 100$$

where V is the volume of the aggregate in millilitres (ml) and W is the mass of the aggregate in grammes.

Problem 4. The relative density of a gravel aggregate is 2.6 . 2000 g of wet aggregate displaced 840 ml of water in a siphon can. Determine the moisture content.

$$\text{moisture content} = \frac{G_s V - W}{G_s (W - V)} \times 100$$

$$= \frac{2.6 \times 840 - 2000}{2.6 (2000 - 840)} \times 100 = 6.1\%$$

It is important to allow for the moisture content of the aggregates when materials are batched by weight for a concrete mix.

> *Problem 5.* A concrete mix requires the following quantities of materials:
> coarse aggregate   640 kg   cement   160 kg
> fine aggregate   320 kg   water   80 kg
> The fine and coarse aggregates on site have moisture contents of 5% and 3% respectively. Calculate the correct batch masses of the aggregates and water.

The definition of moisture content is:

$$\text{moisture content} = \frac{W - W_s}{W_s} \times 100, \text{ where } W \text{ is wet mass and } W_s \text{ is the dry mass.}$$

For the coarse aggregate the dry mass is to be 640 kg and its moisture content is 3%. Using the above formula to find the wet mass:

$$3 = \frac{(W - 640)}{640} \times 100$$

$$W - 640 = \frac{3 \times 640}{100} = 19.2$$

$$W = 640 + 19.2 = \mathbf{659.2 \text{ kg}}$$

Hence if 659.2 kg of damp aggregate are taken it will contain 640 kg of aggregate and 19.2 kg of water.

Similarly for the fine aggregate the dry mass is to be 320 kg and the moisture content is 5%.

$$5 = \frac{(W - 320)}{320} \times 100$$

$$W - 320 = \frac{320 \times 5}{100}$$

$$W - 320 = 16$$

$$W = 320 + 16 = \mathbf{336 \text{ kg}}$$

If 336 kg of damp fine aggregate are taken it will contain 320 kg of aggregate and 16 kg of water.

The amount of water added to the mixer must be reduced by the amount of water contained in the coarse and fine aggregates.

Water = 100 − 19.2 − 16 = **64.8 kg**

The quantities of materials are:
coarse aggregate   659.2 kg   cement   160 kg
fine aggregate   336 kg   water   64.8 kg

Bulking This is the increase in volume of a given mass of sand caused by water films around the sand particles. The extent of bulking depends upon the moisture content and the fineness of the sand, and may be as high as 20–30% at moisture contents about 5–8%. *Fig 5* illustrates the typical variation in volume with moisture content. It will be

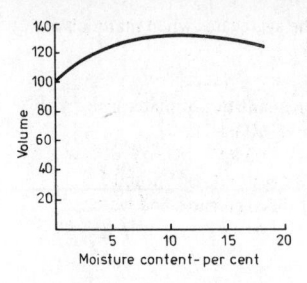

Fig 5 Bulking of a sand

found that the volume of saturated sand is the same as that of dry sand. The bulking can thus be found by measuring the volume of a quantity of damp sand then finding the volume of the sand when it is flooded with water and stirred to expel air bubbles.

Bulking of sand is important when materials are batched by volume since if no allowance is made for bulking the mix will be deficient in sand.

Problem 6. Damp sand is placed in a measuring cylinder and has a volume of 500 ml. When flooded with water the volume of the sand is found to be 420 ml. Find the percentage bulking. A mortar mix is required to contain 0.03 m³ of sand. Find the volume of damp sand that should be used.

The bulking is usually expressed in terms of the saturated volume. Thus the increase in volume of the damp sand is 80 ml.

The bulking = $\frac{80}{420} \times 100\% = 19\%$

To obtain the correct volume of sand in the mix allowing for the bulking the amount of sand must be increased by 19%.

Thus volume of sand = $0.03 \times \frac{119}{100} = 0.0357$ m³

Lightweight concretes Lightweight concretes have densities between 400 kg/m³ and 1800 kg/m³ compared with around 2400 kg/m³ for ordinary concrete. Lightweight concretes have the following advantages: improved thermal insulation; improved fire resistance; reduced loads on foundations and supporting structures.

One method of producing a lightweight concrete is to use lightweight aggregates. The following are some of the lightweight aggregates:

(i) exfoliated vermiculite: used for low strength, insulating concrete.
(ii) expanded clay, shale and slate: used for producing concretes with densities in the range 1400 to 1800 kg/m³. Suitable for structural concrete and light-weight blocks.
(iii) sintered pulverised fuel ash: used as in (ii) above;
(iv) foamed slag: used as (ii) above.
(v) graded wood particles: used for making blocks.

Other methods of producing lightweight concrete are no fines concrete and aerated concrete. No fines concrete contains only coarse aggregate, either natural or lightweight resulting in a honeycombed structure having good insulating properties. In producing aerated concrete bubbles of gas are trapped in the structure which usually only contains fine aggregate. The bubbles can be produced by the use of aluminium powder or by the use of a stable foam. The material is usually autoclaved to reduce the shrinkage and moisture movement.

High-density concretes These have been mainly used for radiation shielding. A number of dense aggregates are possible including:

barytes (barium sulphate); steel shot, steel punchings and sheared steel bars;
iron ore; iron phosphides.

Concrete

Ordinary dense concrete is made from a mixture of coarse aggregate, fine aggregate, cement and water. In order to produce a concrete of the correct strength and durability the proportions of the materials must be carefully selected and controlled on site. The aim is to produce a dense non-porous material. To this end the proportion of fine aggregate to coarse aggregate must be chosen so that when they are combined the fine aggregate will fill the voids spaces in the coarse aggregate. The amount of water added is required to hydrate the cement and to give the mix suitable workability for placing on site.

In discussing concrete mixes the following ratios are helpful:

$$\text{water-cement ratio} = \frac{\text{mass of water}}{\text{mass of cement}}$$

$$\text{aggregate-cement ratio} = \frac{\text{mass of aggregate}}{\text{mass of cement}}$$

Problem 7. A concrete mix is to be made with 50 kg of cement and having a water-cement ratio of 0.45. Find the amount of water to be added to this mix.

Using the above definition of water-cement ratio:

$$0.45 = \frac{\text{mass of water}}{50}$$

mass of water = 0.45 × 50 = 22.5 kg

Problem 8. A concrete mix contains 960 kg of aggregate and 160 kg of cement. Determine the aggregate-cement ratio.

$$\text{aggregate-cement ratio} = \frac{\text{mass of aggregate}}{\text{mass of cement}} = \frac{960}{160} = 6$$

Problem 9. A concrete mix is to be made with the following proportions by mass

1	:	2	:	4	/	0.5
cement	:	fine aggregate	:	coarse aggregate	/	water-cement ratio

The density of the wet concrete is 2400 kg/m^3. Calculate the batch quantities for a 0.25 m^3 batch of wet concrete.

Since the density is 2400 kg/m³ a 0.25 m³ batch will require 2400 × 0.25 = 600 kg of materials. This amount of material must be split in the proportions:

0.5 :	1 :	2 :	4
water :	cement :	fine aggregate :	coarse aggregate

Note that the total of these proportions is:

$0.5 + 1 + 2 + 4 = 7.5$

Mass of water = $\dfrac{0.5}{7.5} \times 600$ = **40 kg**

Mass of cement = $\dfrac{1}{7.5} \times 600$ = **80 kg**

Mass of fine aggregate = $\dfrac{2}{7.5} \times 600$ = **160 kg**

Mass of coarse aggregate = $\dfrac{4}{7.5} \times 600$ = **320 kg**

PROPERTIES OF FRESH CONCRETE

The most important property of fresh concrete is its workability since this relates to the ease with which it can be placed and compacted. Good compaction is essential otherwise voids will result causing a loss of strength. As a simple rule of thumb, 1% voids will cause a 5% loss of strength. There are three tests in use for assessing workability.

Slump test The slump cone and tamping rod are shown in *Fig 6*. The cone is placed on a level, smooth surface and filled in four layers. Each layer is tamped 25 times using the rounded end of the steel rod. The top is struck level with a trowel. During the filling operation the mould is firmly held by standing on the footrests. The cone is then lifted vertically allowing the concrete to slump. By inverting the cone, placing the rod

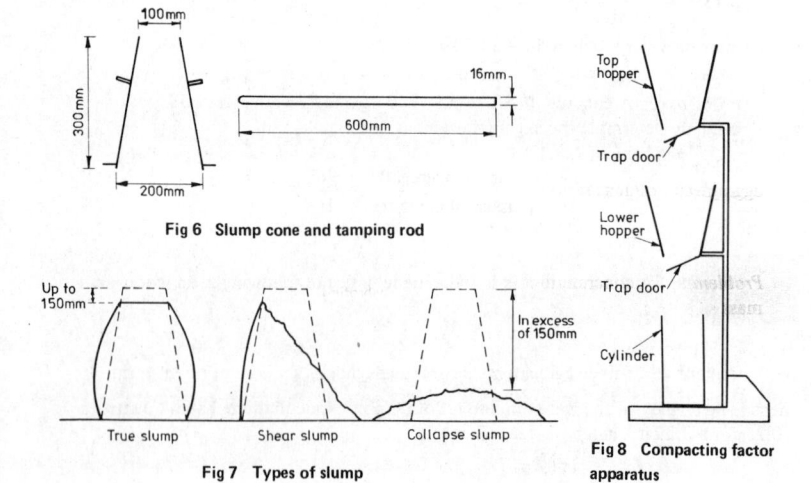

Fig 6 Slump cone and tamping rod

Fig 7 Types of slump

Fig 8 Compacting factor apparatus

across it, the slump of the concrete, to its highest point, may be measured to the nearest 5 mm.

Dry mixes may have no slump so that the test will not show variations in workability. Rich mixes behave satisfactorily. Lean mixes may produce a shear slump in which one half of the cone slides down as shown in *Fig 7*. If shear slump persists when the test is repeated this indicates a lack of cohesion in the mix.

Full details of the method of conducting the slump test are given in BS 1881:1970. The slump test is used to ensure that the workability of a given concrete is maintained at the same level throughout production.

Compacting factor test This consists of two conical hoppers mounted on a frame and a metal cylinder as shown in *Fig 8*. Each of the hoppers has a trap door on the bottom. The following is a very brief outline of the test; full details will be found in BS 1881. The doors on both hoppers are closed and the top is filled with concrete which is placed gently to avoid compaction. The door of the top hopper is opened and the concrete falls into the second hopper. The door of the lower hopper is opened and the concrete allowed to fall into the cylinder. Excess concrete is struck off using two floats slid across the top of the cylinder. The mass of partially compacted concrete in the cylinder is found. By filling the cylinder with fully compacted concrete the mass of fully compacted concrete can be found. The compacting factor is defined by:

$$\text{compacting factor} = \frac{\text{mass of partially compacted concrete}}{\text{mass of fully compacted concrete}}$$

Problem 10. In carrying out a compacting factor test the following results were found:
mass of cylinder = 6.5 kg
mass of cylinder + partially compacted concrete = 18.75 kg
mass of cylinder + fully compacted concrete = 19.9 kg
Calculate the compacting factor.

$$\text{Compacting factor} = \frac{\text{mass of partially compacted concrete}}{\text{mass of fully compacted concrete}}$$

$$= \frac{18.75 - 6.5}{19.9 - 6.5} = 0.91$$

V-B consistometer test Full details of this test may be found in BS 1881, the following being only a brief outline. *Fig 9* shows the details of the equipment. With the aid of the filling hopper the slump cone is filled in the usual manner and then removed. The slump may be measured if required. The transparent plate is then rested on top of the

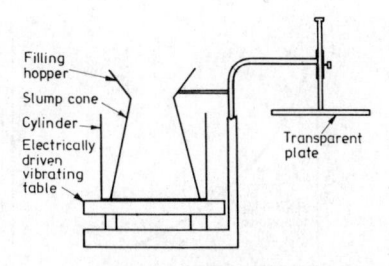

Fig 9 V-B consistometer

concrete and the vibrating table is switched on. The concrete remoulds under vibration and the transparent plate falls. The time is measured between the start of the test and the time when the concrete completely covers the underside of the transparent plate.

The V-B test is very good for dry mixes and has the advantage of measuring the workability under vibration which is related to the method of placing in practice.

It is not possible to obtain a universal correlation between the three tests since the correlation depends on the mix proportions and aggregate properties. The table indicates the likely correlation between the slump test and the V-B test.

Slump (mm)	0–10	10–30	30–60	60–180
V-B (seconds)	>12	6–12	3–6	0–3

Factors affecting workability The main factor affecting the workability is the water content of the mix in kilograms of water per cubic metre of concrete. For a given level of workability the amount of water required will depend on the maximum size of the aggregate, the shape of the aggregate particles and also on their surface texture. The amount of water required per cubic metre for a given workability will decrease as the maximum aggregate size increases, will be less for uncrushed aggregates than for crushed angular aggregates and will be less for aggregates with a smooth surface texture. For a given water-cement ratio with a given aggregate the workability is approximately independent of the aggregate-cement ratio for normal concretes.

HARDENED CONCRETE

Factors affecting the strength of hardened concrete The following factors affect the strength of hardened concrete:
(a) water-cement ratio;
(b) compaction;
(c) aggregate-cement ratio;
(d) surface texture, shape and strength of aggregate particles;
(e) maximum size of aggregate.

By far the most important of these is the water-cement ratio.

The higher the water cement ratio the lower the strength as shown in *Fig 10*. At low water-cement ratios there may be a loss of strength since the workability is so low that adequate compaction may not be achieved. The workability of a mix should be such

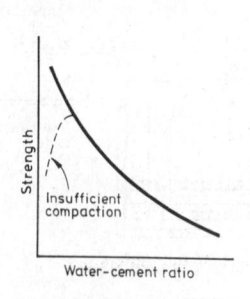

Fig 10 Variation of strength with water - cement ratio

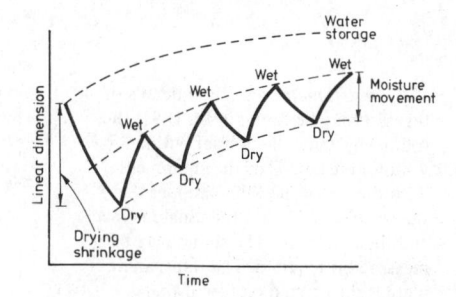

Fig 11 Drying shrinkage and moisture movement of concrete

that compaction can be achieved on site with the method employed since a one per cent of voids will lead to a 5% loss of strength.

Compressive strength test on concrete cubes This test is employed to control the quality of concrete being produced. Since this is a quality control test it is essential that the cubes are made, cured and tested in a standard manner to avoid variations due to the procedure used. The procedure is standardised in BS 1881. The following is only a brief outline of this procedure.

The cubes are made in moulds, usually 150 mm cube, the mould is filled in three layers, each layer being compacted by not less than 35 strokes of a 25 mm square standard tamping bar. Excess concrete is then removed and the top surface trowelled level. The cube is stored for 24 hours at a temperature of 18 to $22°$ C at a relative humidity of not less than 90 per cent. The cube is then demoulded and cured in water at 19 to $21°$ C. When the cube is tested in the compression machine it is placed with the cast faces in contact with the platens of the machine and loaded at the specified rate until failure occurs.

The compressive strength is then calculated as:

$$\text{compressive strength} = \frac{\text{load at failure}}{\text{cross sectional area of cube}}$$

The compressive strength of the site concrete may well be different from the strength of the cube for the following reasons:
(i) different degree of compaction;
(ii) different curing conditions;
(iii) different dimensions between the cube and the structure;
(iv) different methods of load application.

Drying shrinkage and moisture movement When hardened concrete initially dries out there is a decrease in size. If the concrete is again saturated the original size is not fully restored as shown in *Fig 11*. This initial decrease is called the drying shrinkage. Subsequent cycles of wetting and drying will cause alternate increase and decrease of size. This change in dimension is termed the moisture movement and decreases slightly during succeeding cycles of wetting and drying.

Drying shrinkage is associated with the removal of water from the hydrated cement compounds and is restrained by the aggregate. Thus rich mixes have a higher drying shrinkage than mixes with a high proportion of aggregate.

Sulphate attack on hardened concrete Soluble sulphates in the soil, groundwater or seawater, react with the hydrates of tricalcium aluminate to produce calcium sulphoaluminates. This is accompanied by a large increase in volume resulting in the disruption of the concrete. All Portland cements contain tricalcium aluminate and their resistance to sulphate attack is increased by reducing the amount of this compound. The reader is advised to consult the section on sulphate resisting Portland cement earlier in this chapter (see page 67).

Sodium and magnesium sulphates also react with the calcium hydroxide, $Ca(OH)_2$, produced by the hydration of the calcium silicates (see section on hydration of cements). The end product of the reaction is calcium sulphate dihydrate, again accompanied by an expansion of volume.

In addition to the above reactions magnesium sulphate also reacts with the hydrated calcium silicates to give calcium sulphate dihydrate. Thus attack by magnesium sulphate is more serious than that by other sulphates. Sulphate attack is reduced by having dense concrete of low permeability which is achieved with fairly rich mixes of low water-cement ratio. Good compaction is of course essential.

For different types of cement the sulphate resistance increases in the following order:
(i) ordinary and rapid hardening Portland cement;
(ii) low heat Portland cement and Portland blast furnace cement;
(iii) sulphate resisting Portland cement;
(iv) supersulphated cement; (v) high alumina cement.

Efflorescence on concrete Deposits of salts on the surface of concrete is termed efflorescence, it occurs when water containing these salts evaporates at the surface. Efflorescence can be caused by any of the following methods:
(i) the leaching of $Ca(OH)_2$ which reacts with carbon dioxide in the air to give calcium carbonate;
(ii) the use of unwashed seashore aggregates;
(iii) aggregates containing soluble salts.

In the case of concrete exposed to sea water crystallisation of salts in the pores of the concrete may result in disruption of the concrete.

Exercises (*Answers on page 108*)

1 List, and state the formulae, of the main compounds in a Portland cement.

2 Show schematically the hydration reactions of dicalcium silicate and tricalcium silicate.

3 State which compound gives rise to the early strength development of a Portland cement.

4 Referring to *Problem 1*, a cement C has 30% C_3S, 46% C_2S, 5% C_3A and 13% C_4AF. Deduce the likely properties of this cement and state the probable cement type.

5 A sieve analysis was performed on 2 kg of a 20 mm to 5 mm graded coarse aggregate. The table shows the mass of material retained on each sieve.

Sieve size (mm)	37.5	20.0	10.0	5.0
Mass retained (kg)	0	0.4	0.6	0.9

0.1 kg of material passed the 5.0 mm sieve. Determine whether this aggregate complies with BS 882 for 20 mm to 5 mm graded coarse aggregate.

6 A sieve analysis on 200 grammes of fine aggregate gave the following results:

Sieve size (mm)	10.0	5.00	2.36	1.18	600 μm	300 μm	150 μm
Mass retained (g)	0	10	44	16	44	56	20

10 g passed the 150 μm sieve. Determine the grading zone of the fine aggregate.

7 List four types of (a) natural aggregate (b) lightweight aggregate (c) dense aggregate.

8 In a field settling test the height of the silt and clay layer was 4 mm and the height of the sand layer was 75 mm. Is this sand likely to be suitable for concreting from the point of view of silt and clay?

9 When 2000 grammes of dry coarse aggregate were placed in a siphon can the total volume was found to be 755 ml. When 2000 grammes of the same aggregate when damp were placed in the siphon can the total volume displaced was 805 ml. Determine the relative density of the aggregate and the moisture content of the damp aggregate.

10 A concrete mix is to be made with the following quantities of materials: water 45 kg; cement 100 kg; fine aggregate 150 kg; coarse aggregate 350 kg. The fine and coarse aggregates on site have moisture contents of 6% and 3% respectively. Calculate the correct batch masses of the damp aggregates and water.

11 Damp sand in a measuring cylinder had a volume of 750 ml. When flooded with water the volume of the sand was 595 ml. Determine the percentage bulking.

12 For the sand in Exercise 11: a mix, to be batched by volume, required 0.05 m^3 of dry sand. Find the volume of the damp sand that should be used.

13 State three possible advantages of lightweight concrete.

14 A concrete mix is to be made with 100 kg of cement and is to have a water-cement ratio of 0.55 and an aggregate-cement ratio of 5. Find the masses of the water and aggregate in the mix.

15 The quantities of materials for 1 m^3 of a concrete mix are: water 190 kg; cement 385 kg; aggregate 1858 kg. Find the water-cement ratio and the aggregate-cement ratio.

16 A concrete mix is to be made with the following proportions by mass:
    1   :   1.7   :   3.9   /   0.48
    cement : fine aggregate : coarse aggregate / water-cement ratio
The wet density of the concrete is calculated to be 2450 kg/m^3. Determine the quantities of materials required for 0.5 m^3 of wet concrete.

17 List the main factors affecting the workability of a concrete mix.

18 List four reasons why the compressive strength of standard cubes may differ from the strength of the same concrete in a structure.

In the following questions select the correct options.

19 Rapid hardening Portland cement is likely to contain:
    (a) more C_3S than C_2S; (b) more C_2S than C_3S; (c) more C_3A than C_3S;
    (d) more C_3S than C_3A.

20 Sulphate resisting Portland cement is made such that:
    (a) $C_3A < 3.5\%$; (b) $C_3A > 3.5\%$; (c) $C_3S < 3.5\%$; (d) $C_2S < 3.5\%$.

21 Low heat Portland cement would be used:
    (a) where a rapid gain of strength was required;
    (b) when frost was expected;
    (c) in mass concrete work for a water retaining structure;
    (d) for small, heavily reinforced, concrete sections.

22 High alumina cement would be used:
   (a) where large sections are cast;
   (b) where high early strength is required;
   (c) where a very rapid set is required;
   (d) when the temperature exceeds 25° C with a high humidity.

23 River gravels are:
   (a) rounded; (b) angular; (c) flaky; (d) elongated.

24 Crushed limestone aggregates are:
   (a) rounded; (b) angular; (c) irregular; (d) flaky.

25 A concrete is compacted so that 3% of air voids are left. The inherent strength will:
   (a) be increased by about 15%;
   (b) be increased by about 5%;
   (c) be decreased by about 15%;
   (d) be decreased by about 3%.

26 The main use of the slump test is:
   (a) to check the workability of dry mixes;
   (b) to control the workability of a fairly rich mix throughout production;
   (c) check that the strength will be satisfactory;
   (d) measure the moisture content of the aggregates.

27 A concrete mix has a slump of 25 mm. The expected V-B time for this mix would be:
   (a) in excess of 12 s; (b) between 6 and 12 s; (c) between 3 and 6 s; (d) less than 3 s.

28 Which of the following mixes, using the same aggregates is likely to produce the highest strength:
   (a) water 110 kg; cement 200 kg; aggregate 1000 kg.
   (b) water 50 kg; cement 100 kg; aggregate 500 kg.
   (c) water 250 kg; cement 400 kg; aggregate 2000 kg.
   (d) water 20 kg; cement 50 kg; aggregate 250 kg.

29 Concrete has been produced for several days on a contract with a slump between 30 and 50 mm. Following overnight rain the slump of the first mix of the day was 125 mm. This could be due to:
   (a) aggregate stockpiles inadequately protected from rain;
   (b) too much cement added to the mix;
   (c) too short a mixing time;
   (d) a human error when the water was added to the mix.

6 Plastics and paints

Plastics

CHAIN MOLECULES

Man-made plastics are members of the vast family of compounds formed by the atom carbon. For diagrammatic purposes the carbon atom can be imagined as a small ball with four valence arms as shown in *Fig 1(a)*. Each valence arm can grasp the valence arm of another atom.

The simplest atom with which the carbon atom can combine is hydrogen. This has one valence arm so that one carbon atom can combine with four hydrogen atoms (*Fig 1(b)*). The compound so formed is the gas **methane**, a major component of natural gas. As can be seen from the diagram, the chemical formula can be written as CH_4.

The chain structure which distinguishes carbon compounds arises when carbon atoms link with each other. *Fig 2(a)* illustrates the simplest case of the gas **ethane**. *Fig 2(b)* shows the next member of the series, called **propane**. This process of linking can

Fig 1 (above left)   (a) Valency of the carbon atom; (b) methane molecule
Fig 2 (below left)   (a) Ethane molecule; (b) propane molecule
Fig 3 (above centre)   (a) Ethylene molecule; (b) acetylene molecule
Fig 4 (above right)   Benzene ring structure

continue to produce very long molecules, the properties of which change as the length of the chain increases. To start with the materials are gases and as the chain length increases they become liquids and finally waxy solids. These materials are called hydrocarbons and this particular series are called paraffins.

MULTIPLE BONDING

In many compounds the carbon atom's valence arms double up. That is, two atoms may use two or even three valence arms to combine with each other. *Fig 3(a)* illustrates a simple double bond in the gas ethylene and *Fig 3(b)* illustrates a triple bond in the gas acetylene.

These multiple bonds are less stable than the single bond since the additional bonds are strained and thus more reactive. The double bond is of major importance in the manufacture of plastics, paints and adhesives.

RING STRUCTURE

The linear chain molecules are not the only way for carbon and hydrogen to combine. *Fig 4* illustrates a benzene ring. The benzene ring characterises a group of compounds termed the aromatics because of their odours.

MAN-MADE PLASTICS AND POLYMERISATION

Although many long chain molecules exist in natural materials, such as timber, it is only comparatively recently that man has been able to start with simple molecules and join them together.

The simple molecule is called the monomer and the resulting long chain molecule is called a polymer. The essential property of all monomers is that they should contain a double bond. Referring again to *Fig 3(a)* the simplest material with a double bond is ethylene. If the double bond can be broken it will leave two free valence arms as shown

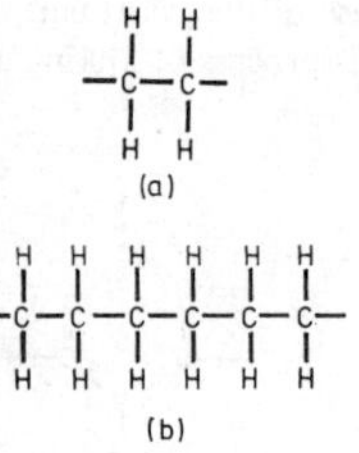

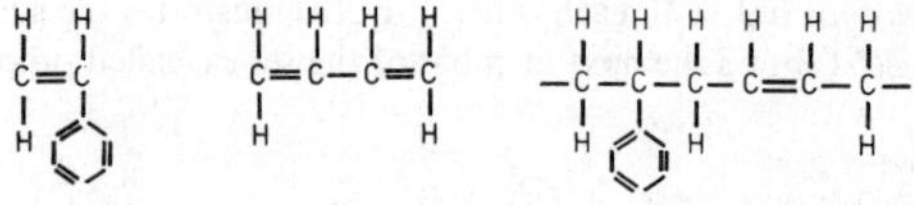

Fig 5 (left) (a) Ethylene molecule with double bond broken; (b) polyethylene

Fig 6 (above) (a) Styrene molecule; (b) butadiene molecule; (c) styrene–butadiene copolymer

in *Fig 5(a)*. If a large number of such molecules exist they can be joined in a long chain as in *Fig 5(b)* and the polymer polyethylene, commonly called polythene, is produced. In producing polythene only one type of molecule is used, such polymers are referred to as homopolymers.

If two or more different molecules are combined the resulting material is called a copolymer. *Fig 6(a)* illustrates a molecule of styrene and *Fig 6(b)* illustrates a molecule of butadiene the resulting copolymer, styrene-butadiene is shown in *Fig 6(c)*. Note that all the atoms on the benzene ring in the styrene have been omitted for simplicity.

The polymerisation of ethylene referred to above illustrates addition polymerisation in which no atoms are gained or lost. Another form of polymerisation in which a small molecule is eliminated is called condensation polymerisation. The small molecule eliminated is often water.

THREE DIMENSIONAL STRUCTURE OF POLYMERS

In the preceding discussion the valence arms of the carbon atom have been illustrated in a two dimensional way. In reality the valence arms are arranged symmetrically in space and thus point to the four vertices of a tetrahedron.

The three dimensional nature of the polymer chain will influence how closely the chains can be packed together in the solid material and will thus control the density of the material.

THERMOPLASTIC AND THERMOSETTING PLASTICS

In the thermoplastic materials the polymer chains are not linked to each other and thus cycles of heating and cooling will repeatedly soften and harden the material. In manufacturing articles from these plastics they are softened by heating and then extruded or rolled.

The polymer chains in thermosetting plastics are cross linked as illustrated in *Fig 7*. These linkages are strong chemical bonds which are created when the plastic article is manufactured. This rigid structure does not soften on heating. Thermosetting materials are suited for moulding into articles requiring some rigidity and heat resistance.

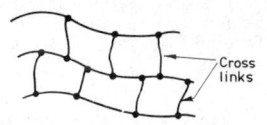

Fig 7 Cross linked polymer chains

PROPERTIES OF PLASTICS

Although the plastics are a very diverse group of materials some general comments on their properties will give an appreciation of the possible limitations of their use. The selection of an incorrect material for a particular application will lead to defects at a later stage.

The plastics have a low Young's modulus of elasticity when compared to the metals. For the metals used in construction the values range from 70 kN/mm^2 for aluminium to 210 kN/mm^2 for steel. The values of Young's modulus for the majority of plastics lie in the range 1.5 to 11 kN/mm^2. Thus plastic materials will show large deflections unless suitable precautions are taken.

The density of plastics is generally low, for instance the density of polythene is slightly less than that of water and the density of polyvinylchloride is slightly greater than that of water. Many plastics can be produced as foams of very low density suitable as insulation materials.

The softening point of the thermoplastic materials is of consequence in a number of applications. The softening points range from 75° C for low density polythene to 300° C for polytetrafluoroethylene.

The coefficient of expansion, particularly of many of the thermoplastics is high compared with other constructional materials. For instance the coefficient of expansion of polyvinyl chloride is about five times that of steel. Thus if plastics are used in situations where temperature changes will occur, for instance in domestic waste systems or gutters, provision must be made for expansion. Behaviour in event of fire is important. The thermoplastics will soften. Many of the plastics, such as polythene, burn. The thermosetting plastics, although they do not soften, will eventually decompose.

Weathering effects are caused primarily by ultra-violet radiation, temperature, water and oxygen. Ultraviolet radiation can cause discoloration and in some cases brittleness by creating a change in the chemical structure. Plastics for exterior use, that are susceptible to these changes, usually incorporate a suitable ultra-violet absorber. Some polymers, for instance, polyvinyl chloride may become brittle at low temperatures.

Degradation by oxidation can be accelerated by high temperatures, ultra-violet radiation and moisture. The molecular structure of the polymer is changed with a consequent loss of strength.

The electrical resistance of the plastics is very high and many applications for electrical components and insulation exist.

TYPES OF THERMOPLASTICS

The following paragraphs give a brief outline of some of the thermoplastics used in construction.

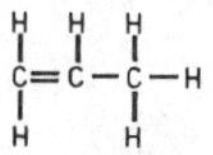

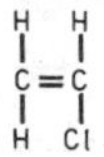

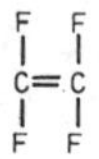

$$\begin{array}{ccc} H & H & H \\ | & | & | \\ C=C-C-H \\ | & | & | \\ H & & H \end{array}$$

$$\begin{array}{cc} H & H \\ | & | \\ C=C \\ | & | \\ H & Cl \end{array}$$

$$\begin{array}{cc} F & F \\ | & | \\ C=C \\ | & | \\ F & F \end{array}$$

Fig 8 (left)   Propylene molecule

Fig 9 (centre)   Vinyl chloride molecule

Fig 10 (right)   Tetrafluoroethylene molecule

Polyethylene (Polythene)   The polymerisation of ethylene has already been discussed. There exists two forms of polythene: low density polythene, in which the polymer chains have side branches resulting in a low density packing, and high density polythene where the polymer chains are linear. Polythene is a translucent, waxy material which burns vigorously in the same manner as a candle. It has a very high coefficient of expansion and a poor resistance to ultra-violet light unless carbon black is incorporated.

Uses

sheeting for damp-proof membranes and vapour barriers;
cold water tanks and cisterns;
cold water pipes.

Polypropylene   Formed by the polymerisation of propylene, the structure of which is shown in *Fig 8*. It is similar to polythene but has a higher softening temperature and greater rigidity. Becomes brittle below $0°$ C. Polypropylene can be drawn into a fibre to give string. Suitable forms of this fibre are used to reinforce concrete pile shells to improve their impact resistance.

Polyvinyl Chloride (p.v.c.)   The monomer is shown in *Fig 9* where the chlorine atom is denoted by Cl. Polymerisation leads to a structure similar to polythene but with one in four hydrogen atoms replaced by a chlorine atom. P.V.C. is a rigid transparent material, however it may be rendered flexible by the addition of plasticisers. The properties of p.v.c. can be varied widely by incorporating fillers, stabilisers and plasticisers. The softening point is $80°$ C but it is self extinguishing in event of fire.

Uses:

The uses of p.v.c. are numerous, the following being only a few:
guttering, down pipes and waste systems;
pipes, which can be solvent welded, for cold water systems;
floor tiles with suitable fillers;
flexible transparent sheet;

flexible flooring material;
flexible coatings on metals and timber;
cable insulation;
water stops;
transparent corrugated sheets;
expanded p.v.c. for heat insulation.

As long as the p.v.c. is suitably formulated it has good weathering properties although ultra-violet light can seriously reduce the light transmission properties of the transparent sheet material.

By replacing some of the hydrogen atoms in p.v.c. by chlorine atoms chlorinated p.v.c. is produced which has a higher softening point than p.v.c.

Polyvinyl Fluoride (p.v.f.) This plastic is similar to p.v.c. but a fluorine atom replaces the chlorine atom. P.V.F. has an excellent weather resistance and is used for coating metal cladding panels.

Polytetrafluoroethylene (p.t.f.e.) The monomer is shown in *Fig 10*. The polymer has a high melting point and a very low coefficient of friction.

Uses:
lining cooking pans;
bridge bearings;
as a tape for sealing pipe joints.

Polystyrene The monomer, styrene, was seen in *Fig 6(a)*. Polystyrene is a clear, brittle plastic at room temperatures. It burns readily. Its main use in building construction is as expanded polystyrene as a thermal insulation material.

Acrylic Plastics The best known of these is polymethyl methacrylate, perhaps better known under the trade name 'Perspex'. It is a rigid transparent plastic and has thus found use as roof lights where its good weathering resistance is valuable. Other uses include lighting control diffusers for luminaires and also baths.

Acrylonitrile Butadiene Styrene (a.b.s.) A.B.S. is a rigid, tough plastic, the properties of which can be modified by altering the proportions of the three monomers in the co-polymer. It is used for pipes where its higher softening point than p.v.c. is an advantage. Its other applications are as diverse as rain water goods, refrigerator parts, parts for dishwashing machines, telephones and heels for shoes.

Polycarbonate A tough, rigid, transparent plastic which is used for shatterproof or vandal-proof glazing.

Nylon The nylons constitute a family of plastics with diverse properties. They are generally tough materials, with fairly high melting points and a low coefficient of friction. Applications include ropes, pipes, catches and sliding door parts.

Polyurethanes A wide family some of which are thermoplastic and some thermo-setting. They are often used as foams where according to the formulation they may be rigid or flexible. The foams have been used for flexible pipe insulation, furniture upholstery and in rigid forms as part of composite partition panels. There is concern over the use of foams in furniture due to the rate of burning, smoke produced and the production of toxic fumes in the event of fire.

TYPES OF THERMOSETTING PLASTICS

Phenol Formaldehyde This is made by the condensation polymerisation of phenol and formaldehyde and produces a crosslinked structure which is thermosetting. It is a hard, durable plastic with good electrical insulation properties. The colour of the plastic is brown to black. It is used for electrical components, timber adhesives and the production of laminates. The laminates usually have a decorative finish applied to them.

Urea Formaldehyde and Melamine Formaldehyde These are similar to phenol formaldehyde in their properties but are light coloured. The uses are also similar to those of phenol formaldehyde. Melamine formaldehyde is often used as decorative coating for laminates. Foamed urea formaldehyde has been widely used as a cavity fill material for thermal insulation.

Epoxy Resins The chemistry of this family is complex. They can be crosslinked by the addition of a hardener. The uses include flooring materials, paints and adhesives.

Polyester Resins The major use of the polyester resins in building construction is as glass-fibre reinforced polyester resin (g.r.p.), which has found applications ranging from water tanks, translucent corrugated sheeting, swimming pools to childrens' playground equipment. The weathering characteristics of g.r.p. depend upon the careful selection of the polyester resin and the method of manufacture. Deterioration of g.r.p. depends on the failure of a thin layer of resin which covers the fibres near the surface. If the fibres are exposed, the surface roughens, collects dirt and the transparency is seriously reduced.

Paints

Paint technology is a complex subject and the following is a very brief outline. The functions of a paint are to protect the substrate, to provide a durable surface finish and to provide a decorative finish. The emphasis to be placed on each of these factors will depend on the circumstances involved. For simplicity paints for building will be divided into oil paints and emulsion paints.

OIL PAINT SYSTEMS

The oil paint system is applied in several stages, each having its own particular function. Primers which are specifically formulated for each type of substrate must have good adhesion to the substrate and may provide corrosion protection. The coat must be such that good adhesion of subsequent coats is obtained. Undercoats have a high pigment content to give good opacity and normally have a matt finish to ensure good adhesion of subsequent coats. The finishing coat provides a surface of suitable durability and aesthetic value.

COMPOSITION OF OIL PAINTS

An oil paint has a liquid and a solid component. The liquid component which is termed a binder, vehicle, film former or medium will convert the liquid coating to a dry solid film having suitable elasticity, adhesion and other properties required from the paint. The solid component consisting of pigments and extenders will provide the opacity, the colour and in some cases the corrosion protection to substrate.

PAINT VEHICLES

These are often based on materials which will dry in the presence of oxygen. They may be natural drying oils and resins, for instance linseed oil, soya bean oil, tung oil and copal resin or synthetic resins such as alkyd, polyurethane, epoxy and phenolic.

In many decorative paints the natural oils are reacted with a synthetic resin to give the required properties. The oil modified alkyd resins, containing over 60 per cent of a drying oil have good flow, gloss, flexibility and weather resistance but are not resistant to alkalis such as occur in Portland cement products. The air curing, single pack oil modified polyurethanes, often referred to as urethane alkyds, give a harder, tougher and more water resistant film than most of the alkyd paints. Epoxy resins can also be oil modified to give a single pack, air curing epoxy ester paint with improved chemical resistance.

The reaction of an alkyd resin or an oil with a polyamide results in the formation of a thixotropic resin. This can be used by itself or as an additive to other resins in the paint vehicle.

Two pack polyurethane and epoxy resins can be used for specialised purposes where chemical and abrasion resistant surfaces are required. Silicone resins are used as a basis of heat resistant paints and can be copolymerised with alkyd and polyurethane resins to improve their durability, water repellancy and temperature resistance.

The viscosity of the paint must be suitable for the type of application being used in order to possess suitable flow properties. The viscosity is adjusted by the addition of thinners. Many decorative paints can be thinned with white spirit but some paints require specialised thinners. The rate of drying of the paint is controlled by the addition of driers typically the napthenates and octanoates of manganese, cobalt and lead.

PIGMENTS AND EXTENDERS

The function of the pigment is to give the paint covering power, opacity and colour. In some instances the pigment may provide corrosion protection for the substrate. Pigments may be organic or inorganic materials.

Most decorative paints contain a high proportion of a white pigment such as titanium dioxide, zinc oxide or white lead. Titanium dioxide being the most commonly used. In addition colour pigments may be added to give the desired colour, examples are chrome yellow, prussian blue and carbon black.

Calcium plumbate is used in primers for wood and galvanised steel. Zinc dust is used in paints for corrosion protection of steel. Iron oxide pigments are widely used in paints for protecting primed steel structures. Aluminium powder is used in sealing primers and heat resisting paints.

Extenders which have little opacity in the drying oil are added to improve adhesion, gives suitable roughness to undercoats, improve the flow properties, reduce the settlement of heavier pigments or to reduce the price. Examples are barytes, whiting, china clay and Fuller's earth.

DRYING OF A PAINT FILM

Initially the thinners evaporate so that the paint can no longer be brushed. The drying oils for air drying paints all contain double bonds in their structure. These double bonds react with the oxygen in the air which initiates polymerisation between the molecules in the drying oil. The final paint film is thus a cross-linked polymer.

DEFECTS OF PAINT FILMS

Bleaching which is a whitening or a colour change of the paint film can be caused by chemical action on the pigments.

Bleeding which is discolouration is caused by the paint vehicle dissolving some constituent of the surface. An example is the painting of a surface coated with bitumen or creosote.

Blistering of the paint film can occur when water trapped behind the film vapourises. It can also be caused by resin exuding from knots or excessive heat.

Chalking, which is the formation of a powdery surface may result from overthinning the paint.

Cissing in which the paint rolls back leaving small bare patches may occur on greasy or very smooth surfaces.

Cracking and **crazing** can occur when hard finishes are applied over soft coatings or before previous coats are thoroughly dry.

Flaking or **peeling** can arise by painting damp surfaces, powdery or friable surfaces, such as distemper, or lack of adhesion on shiny surfaces. Efflorescence can also cause flaking.

Saponification can arise on alkaline surfaces typified by Portland cement products. The drying oil reacts with the alkali to give a non-drying oily soap.

Wrinkling can occur if the paint is applied too thickly so that the film does not dry evenly throughout.

EMULSION PAINTS

In an emulsion paint the resin is dispersed in water. The paint film dries by evaporation of the water and at the same time the resin particles flow together and coalesce to form a film.

Emulsion paints are water thinnable and are available in matt, eggshell and gloss finish. The film once dry is no longer soluble in water. The resins used are homopolymers or copolymers based on polyvinyl acetate or acrylic resins. Styrene-butadiene is also used. The pigments and extenders are similar to those previously described for gloss paints.

Several other materials are added in order to produce a satisfactory paint these include: thickeners, biocides to inhibit the growth of micro-organisms, rust inhibitors, pigment wetting agents and solvents designed to aid the resin particles to coalesce.

Emulsion paints are formulated for many applications. They may be used on many internal and external surfaces including plaster and renderings. Direct application to bare metals is not generally recommended. Acrylic emulsion primers are used on joinery where rapid drying is essential. Exterior masonry paints, which are water thinnable are based on acrylic or vinyl copolymers reinforced with mica, quartz, granite sand or nylon fibres.

Exercises (*Answers on page 108*)

In the following questions select the correct options.
1. Ethylene is the monomer of
   - (a) p.v.c.
   - (b) polypropylene;
   - (c) p.f.t.e.
   - (d) polythene.

2. The monomer of polypropylene is:
   - (a) ethane;
   - (b) methane;
   - (c) polypropylene;
   - (d) propylene;
   - (e) p.v.c.

3. A copolymer is always obtained by the polymerisation of:
   - (a) the molecules of a single monomer;
   - (b) the molecules of two monomers;
   - (c) the molecules of at least two monomers;
   - (d) by crosslinking of molecular chains.

4. The thermal expansion of plastics is:
   - (a) usually greater than that of metals;
   - (b) usually less than that of metals;
   - (c) about the same as metals.

5. The modulus of elasticity of plastics is:
   - (a) usually greater than that of metals;
   - (b) usually less than that of metals;
   - (c) about the same as metals.

6. Thermoplastic materials are:
   - (a) p.v.c.
   - (b) phenol formaldehyde;
   - (c) p.v.f.
   - (d) epoxy resins;
   - (e) urea formaldehyde;
   - (f) polymethyl methacrylate.

7. Thermosetting materials are:
   - (a) a.b.s.
   - (b) epoxy resins;
   - (c) polythene;
   - (d) p.v.c.
   - (e) urea formaldehyde;
   - (f) polymethyl methacrylate.

8. P.V.C. is suitable for:
   - (a) gutters;
   - (b) hot water systems;
   - (c) shatterproof glazing;
   - (d) transparent corrugated sheets;
   - (e) fire resistant air conditioning ducts.

9. A tape for sealing pipe joints is made of:
   - (a) p.v.c.
   - (b) polythene;
   - (c) polytetrafluoroethylene;
   - (d) a.b.s.

10 Carbon black is added to polythene to:
   (a) increase its density;
   (b) increase its resistance to sunlight;
   (c) alter its colour;
   (d) reduce water penetration when used as a d.p.c.

11 Materials commonly used for pipes in building are:
   (a) a.b.s.
   (b) p.v.f.
   (c) polyvinyl chloride;
   (d) phenol formaldehyde;
   (e) polythene;
   (f) polyurethane;
   (g) p.t.f.e.
   (h) polymethyl methacrylate.

12 Decorative laminates are made from:
   (a) p.v.c.
   (b) phenol formaldehyde;
   (c) polythene;
   (d) a.b.s.

13 The correct order of a paint system is:
   (a) primer, finish, undercoat;
   (b) undercoat, primer, finish;
   (c) primer, undercoat, finish;
   (d) finish, undercoat, primer.

14 A paint based on a drying oil dries by:
   (a) evaporation of the oil;
   (b) evaporation, oxidation and polymerisation.
   (c) heating;
   (d) evaporation of the thinners.

15 Saponification is due to:
   (a) painting a wet timber surface;
   (b) painting an alkaline surface;
   (c) applying too great a coat thickness;
   (d) too much sunlight whilst painting.

16 Emulsion paints are thinned with:
   (a) water;
   (b) white spirit;
   (c) paraffin;
   (d) polyvinyl acetate.

17 Blistering of a paint film is due to:
   (a) painting a wet timber surface;
   (b) painting an alkaline surface;
   (c) adding too much thinners;
   (d) painting a greasy surface.

18 A typical oil paint would have as its main constituents:
   (a) drying oil and extender;
   (b) drying oil, thinners, pigments and extender;
   (c) drying oil, driers, pigments and extender;
   (d) drying oil, thinners, driers, pigments and extender.

19 Emulsion paints are suitable for painting:
   (a) rendering;
   (b) ceilings;
   (c) steel radiators;
   (d) external aluminium cladding;
   (e) brickwork.

20 A suitable primer for galvanised steel is based on:
   (a) zinc oxide;
   (b) prussian blue;
   (c) calcium plumbate;
   (d) whiting.

7 Timber, bricks, blocks and plaster

Timber

GROWTH AND STRUCTURE OF TIMBER

The root system of a tree absorbs water and dissolved salts from the soil. This water, called sap, is conducted by the sapwood to the leaves. The leaves contain chlorophyll and by a process of photosynthesis break up carbon dioxide absorbed by the leaves into carbon and oxygen. The carbon combines with the sap to produce sugars, starch, celluloses and other carbohydrates necessary for the growth of the tree. Some of the liberated oxygen is given off by the leaves. The carbohydrates then return downwards to feed and maintain the tree.

Growth takes place by the fission of cells in the cambium layer and also at the tips of branches. The position of the cambium layer is shown in *Fig 1* which shows a schematic cross-section of the trunk of a tree. The new cells formed provide new sapwood and inner bark. The cambium layer which shows as a ring on the section does in reality encase the complete tree.

Fig 1 Cross-section of a tree trunk

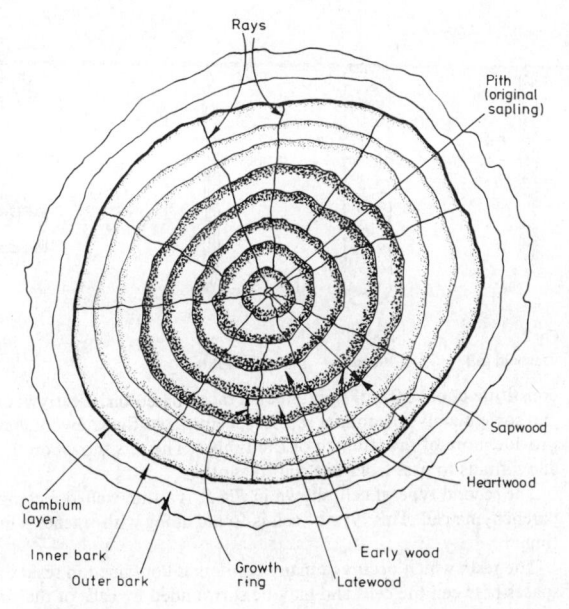

The following explanations relate to *Fig 1*. The main features of the section of a tree are the growth rings each of which represents the timber formed in one growing season, which in temperate climates is a year. The earlywood grows in the early part of the season and is less dense than the latewood which grows towards the end of the season. In some timbers the latewood is darker in colour than the earlywood making the growth rings very distinctive.

The sapwood contains the living cells in the tree. Its function is to conduct sap from the roots to the leaves and to store reserves of food material. The sapwood is permeable to liquids and is less durable than the heartwood. The heartwood contains cells that are no longer living; these do not contain sap. In some timbers the sapwood is of a distinctly lighter colour than the heartwood.

The rays, some of which start at the pith serve the dual purpose of storing food reserves and provide for horizontal transport of the sap. The inner bark serves to convey food from the leaves to the growing parts of the tree. The outer bark protects the tree from damage and provides insulation against the external environment.

STRUCTURE OF SOFTWOODS

All trees are composed of cells; the type and nature of these cells varies between the hardwoods and the softwoods. The softwoods belong to the botanic family of gymnosperms, of which the most important are the conifers. The term softwood does not necessarily relate to the hardness of the timber. Pine is a softwood whereas oak is a hardwood but balsa is also a hardwood.

The conifers have a more primitive structure than the hardwoods being composed of only two types of cell. The first type of cell is called a tracheid and is a long, needle or cigar-shaped cell as shown in *Fig 2*. Tracheids are usually 2-5 mm in length and

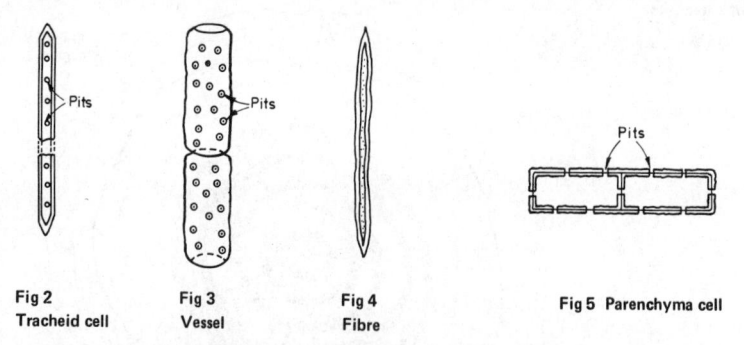

Fig 2
Tracheid cell

Fig 3
Vessel

Fig 4
Fibre

Fig 5 Parenchyma cell

constitute about 90% of the timber. Tracheids from the earlywood have thin walls and serve mainly to conduct sap. Tracheids from the latewood have thicker walls and provide more of the strength of the timber. The pits shown on the walls of the cells allow fluids to pass from one cell to another.

The second type of cell, shown in *Fig 5*, is a thin walled rectangular cell called a parenchyma cell. This type of cell is found along with tracheids in the rays of the timber.

The resin which occurs in many conifers is contained in resin ducts which are large spaces between the cells and may be surrounded by cells of the parenchymal type.

STRUCTURE OF HARDWOODS

The structure of hardwoods is more complicated than that of softwoods. The vessels, as shown in *Fig 3*, conduct the sap. These vessels are formed from a series of cells, the end walls of which allow the passage of sap. The vessel is characteristic of hardwoods and the arrangement of the vessels is a major clue to the identification of a hardwood. The fibre, illustrated in *Fig 4*, is a long narrow cell with thick walls which contributes the majority of the mechanical strength of the timber. Some hardwoods contain tracheids similar to those in a softwood.

The majority of the parenchyma are in the rays which may account for more than a fifth of the total wood. The rays are usually several cells wide compared with the softwoods where the rays are usually only one cell wide. Parenchyma can occur throughout the wood either associated with the vessels or scattered throughout the fibres.

INSECT ATTACK ON TIMBER

Many insects attack timber but the following will be confined to beetles that attack timber in buildings in Great Britain. The life cycles of all wood boring beetles are similar. The eggs are laid by the female beetle usually in cracks or holes in the timber. The larvae which hatch from the eggs bore through the timber extracting their food from the timber. This stage may last several years. The larvae then pupate within the timber and emerge as fully adult beetles. The first evidence of the attack is usually the exit holes left by the beetles.

Common furniture beetle The common furniture beetle (Anobium punctatum) is the best known wood boring beetle and attacks particularly the sapwood of a wide variety of hardwoods and softwoods. A number of tropical hardwoods appear to be immune to attack. Wickerwork and plywood made with animal glues are particularly susceptible. The adult beetles emerge between May and August leaving a circular exit hole 1–2 mm in diameter. The bore dust consists of small ellipsoidal pellets. The beetles are 2.5–5 mm in length and are reddish to blackish-brown. The female lays up to one hundred eggs and the attack can be spread by the beetles crawling or flying to unaffected timber. Attack is favoured by using timber with a high proportion of sapwood and a high moisture content. Temperatures not exceeding 22° C are also favourable. The attack is accelerated in timbers suffering fungal decay.

Death watch beetle The death watch beetle (Xestobium rufo-villosum) attacks hardwoods such as oak and chestnut. The attack usually occurs in damp timber which at some stage has suffered fungal decay. The beetles emerge between the end of March and June leaving a circular exit hole about 3 mm in diameter. The bore dust consists of bun shaped pellets. The beetles are 6–9 mm long, are chocolate brown in colour and covered with patches of yellowish hairs. The beetles produce a tapping sound during the mating season which gives rise to its name. The beetles are rather sluggish and seldom fly. The female beetle usually lays 40 to 60 eggs.

The larvae penetrate deeply into the timber often rendering it structurally useless. The author has removed oak floor joists from an old house which, beneath the surface of the timber, consist of nothing more than a mere skeleton of the original wood. These floor joists had collapsed under the weight of the furniture in the room.

House longhorn beetle The longhorn beetles are essentially forest insects but the house longhorn beetle (Hylotrupes bajulus) attacks seasoned softwoods usually the sapwood. Serious damage has been caused to structural timbers in the south of England. The beetles emerge between July and September leaving oval exit holes having a major diameter of 6–10 mm. The beetles are 10 to 20 mm in length and are usually black but sometimes brown. The larvae, which may be up to 30 mm in length bore parallel to the grain while leaving a skin of sound wood on the surface. These borings may create corrugations on the surface of the timber which may be a useful method of detecting the attack. Under the sound skin of wood the timber may be completely disintegrated. If an attack of house longhorn beetle is detected it should be treated without delay under expert guidance. The activity of this beetle appears to be favoured by warm temperatures.

PREVENTION OF INSECT ATTACK

The use of preservative treated timber for new and replacement work would seem to be the best prevention against attack by insects. The Building Regulations specify areas in which roof timbers for new houses must be treated with preservative to prevent attack by longhorn beetle. There are more preservatives which are capable of preventing an attack than there are for eliminating an infestation. Also, the application of preservatives before construction is easier than after construction. It is essential to ensure that the timber is kept as dry as possible after construction. Regular maintenance of gutters, downpipes, flashings and air vents will ensure that the timber is kept as dry as possible.

Care should be taken to ensure that infested timber is not introduced into the building. Many cases of woodworm attack have been started by storing old, infested furniture in roof spaces. Many attacks by death watch beetles have been started by the use of secondhand oak beams.

TREATMENT OF INSECT ATTACK

The extent of structural damage must be ascertained and where replacement timber is necessary it must be treated with a suitable preservative. The remaining timber can be treated with a suitable preservative either by brushing, low pressure spraying or the use of an emulsion preservative. Injection of preservative into the exit holes can assist penetration of the preservative particularly when the attack is small.

Fumigation, usually with methyl bromide, is an effective treatment for places that can be adequately sealed. Treatment of valuable furniture may be carried out this way. It is not usually a suitable treatment for structural timber in roof spaces since these cannot be made adequately gas tight. Fumigation does not leave a residual toxic deposit in the wood so it does not prevent reinfestation.

The annual use of insectidal smoke treatments which kill the beetles on emergence will eradicate an attack of furniture beetle or death watch beetle over a period of years.

FUNGAL ATTACK ON TIMBER

Fungi are a low form of plant life which obtain their food from the dead or living parts of other plants. Fungi develop fine tubes, called hyphae which branch and form a mycelium. The mycelium may be composed of strands which enable the fungus to spread. Most of the growth of the fungus is inside the wood under attack and the attack may not be obvious until a fruiting body is produced. The fruiting bodies produce millions of spores which will germinate under suitable conditions.

DRY ROT FUNGUS

The dry rot fungus (Merulius lacrymans) is the most virulent of the fungi attacking timber in buildings and a knowledge of its appearance and effects is important. In damp conditions it produces white cotton-wool like masses. In less damp conditions it forms a grey felted skin which may have tinges of lilac and yellow patches. Grey branching strands with a thickness up to that of a pencil may be formed. These strands conduct water enabling the fungus to spread to relatively dry timber. The fungus is capable of penetrating brickwork and passing over other inert materials.

The fruit bodies consist of pancake-like plates or thick broad brackets which have a tough consistence and are white or grey when young. When it is ripe the surface is covered by reddish-brown spores. Drops of moisture are exuded giving rise to the name lacrymans or weeping. The fruit body may be any size up to a metre or more across. The fungus has a distinctive smell which is decidedly unpleasant in a well-established attack.

Wood rotted by dry rot appears a dull brown colour, is light in weight and breaks into rectangular pieces formed by deep cracks along and across the timber. The timber is dry and crumbly thus giving rise to the name 'dry rot'. The conditions under which dry rot thrives are: a moisture content of the timber in excess of 20%; still air conditions and a temperature above freezing but below $27°$ C.

PREVENTION AND ERADICATION OF DRY ROT

Preventive maintenance on an existing building will consist of regular inspection and clearance of all gutters, valleys, rainwater pipes and gullies. Sub-floor vents must be kept unobstructed and soil should not be banked above the damp-proof course. Patches of dampness should be dealt with immediately.

Eradication of dry rot would be done in the following stages:
(a) Determine extent of attack.
(b) Remove all timber which is decayed, cutting away for at least 300 mm beyond the last signs of attack. Burn all timber removed.
(c) Strip off plaster containing strands of the fungus and rake the joints in the brickwork.
(d) The walls should be treated with a fungicide. With solid walls or where the fungus has penetrated deeply into the brickwork irrigation via holes drilled at 225 mm centres may be necessary.
(e) Under suspended floors 150 mm of soil should be removed, the soil treated with fungicide and concrete laid to replace the soil removed.
(f) All remaining timber within two metres of the attack should be treated with preservative.
(g) All replacement timber must be treated with preservative.
(h) The source of dampness must be eliminated.

WET ROT

Wet rot is caused by a number of fungi. The amount of damage caused to floors, roofs and external joinery may be considerable but the eradication is easier than with dry rot.

The **cellar fungus** (Coniophora cerebella) causes the wood to darken and split with mainly longitudinal cracks. Frequently the outside of the wood shows no visible signs of rot. The mycelium consists of dark brown strands which do not exceed in thickness that of fine twine.

The **white pore fungus** (Poria vaillanti) grows in very damp conditions and the wood that is attacked is generally much wetter than that attacked by dry rot. The decayed wood resembles that attacked by dry rot. The mycelium is white and the fruit body consists of a whitish plate pitted with fine pores which does not produce reddish-brown spores.

Both the above fungi die when the source of moisture is removed. Remedial work thus consists of curing the dampness and replacing affected timber with preservative treated timber.

TIMBER PRESERVATIVES

A wide range of preservatives now exist and the manufacturers advice should be sought concerning specific applications. The following paragraphs outline the main types used for treating sound timber. Different formulations may be used as fungicides and insecticides for eradicating fungal and insect attack.

Creosote is the best known of the tar oil preservatives and is suitable for external timbers such as fence posts. Such timber cannot then be painted. A number of tar-oil derivatives are made in a range of colours suitable for decorative treatment of external timber.

The **highly fixed water-borne preservatives** have ingredients which react with the wood to produce insoluble compounds which are thus resistant to leaching. Examples of this type of preservative are those based on copper/chrome/arsenic. They are suitable for vacuum-pressure impreganation of timber.

The **leachable water-borne preservatives** are suitable for the treatment of timber for internal use. Sodium octoborate is applied by the diffusion process to green timber which is then seasoned in the usual way.

The **organic solvent preservatives** may contain copper or zinc napthenates, pentachlorophenol or tri-n-butyltin oxide. The use of an organic solvent gives good penetration and is thus suitable for application by brushing, spraying or dipping. No distortion of the timber occurs so they are suitable for treating joinery. The solvents are expensive compared with water. Precautions must be taken to ensure the safety of operatives and to reduce fire risks when these preservatives are used.

Emulsion preservatives have the consistence of mayonnaise. They can be applied in thick layers to *in situ* timber suffering decay. This allows the preservative to diffuse into the timber over a period of days thus absorbing more preservative than would be possible by brush application.

Bricks

Bricks vary widely in their properties and the properties required of a brick depend upon the use; for example bricks used in a parapet wall, which is exposed to rain and frost, must have a greater durability than bricks used for internal brickwork. Many failures of brickwork are due to the use of an incorrect quality of brick. An understanding of the classification of bricks should lead to the selection of the correct bricks for the application envisaged.

Classification of clay bricks

The classification given in BS 3921:1974 for clay bricks divides bricks according to their variety, their quality and their type.

Varieties
(i) **Common**   for general building work.
(ii) **Facing**   having an attractive appearance.
(iii) **Engineering**   a semi-vitreous brick of defined properties.

Qualities
(i) **Internal quality**
(ii) **Ordinary quality**   normally durable in the external face of a building.
(iii) **Special quality**   durable in extreme exposure; for example in parapet walls, retaining walls or pavings.

Types
(i) **Solid**   holes do not exceed 25% of the volume and frogs do not exceed 20% of the volume.
(ii) **Perforated**   holes exceed 25% of the volume and the area of any one hole shall not exceed 3000 mm^2.
(iii) **Hollow**   holes exceed 25% of the volume
(iv) **Cellular**   holes closed at one end exceed 20% of the volume.
(v) **Specials**

The above paragraphs outline the classification system and the reader is advised to consult BS 3921 for full details. Using the above classification a brick may be described, for example, as a **Solid facing** brick of **Special** quality. Specific requirements exist for the different qualities of brick and also for engineering and load bearing bricks. Certain tests upon the properties of the bricks are required and the reader is advised to consult BS 3921 for methods of conducting these tests.

The average strength is found by determining the average compressive strength of 10 bricks. The average water absorption is determined from a 5 hour boiling test or vacuum absorption test on 10 bricks. The soluble salts analysis determines the percentage soluble sulphate and percentage calcium, magnesium, potassium and sodium present. Sulphates are important because of the liability of sulphate attack on Portland cement in the mortar. The efflorescence test, in which the bricks are saturated and are allowed to dry with only one surface exposed to the air, shows the likelihood of efflorescence occurring in practice.

As a result of this test the efflorescence is classified as nil, slight, moderate, heavy or serious. Moderate implies that up to 50% of the face of the bricks is covered in efflorescence but that there is no breakdown of the surface of the brick.

Qualities and requirements of clay bricks

The qualities and types of clay bricks are discussed in the following paragraphs.

FACING AND COMMON BRICKS OF ORDINARY QUALITY

The finish of these bricks is such that they shall be well fired and reasonably free from deep cracks and damage. When cut they should show a reasonably uniform texture. The average strength should not be less than 5.2 N/mm^2. The efflorescence should not be worse than moderate.

FACING AND COMMON BRICKS OF SPECIAL QUALITY

The strength, efflorescence and finish are similar to those of ordinary quality except that they should be hard fired. The bricks must satisfy one of the following criteria of frost resistance:
(i) the brick has performed satisfactorily for at least three years under conditions of exposure at least as severe as the proposed use.
(ii) the brick has performed satisfactorily for at least three years in a suitable test panel.
(iii) the average strength is not less than 48.5 N/mm^2 or the average water absorbtion is not greater than 7%.

The soluble sulphate must not exceed 0.5% by weight. The calcium must not exceed 0.1%, the manesium, potassium and sodium must each not exceed 0.03%.

BRICKS FOR INTERNAL WALLS

The requirements of these bricks are similar to those of ordinary quality except that no mention is made of the degree of firing. Bricks intended only for non load bearing partitions should have an average strength of not less than 1.4 N/mm^2.

ENGINEERING BRICKS

There are two classes:
Class A requires that the average strength is not less than 69.0 N/mm^2 and the average water absorption is not more than 4.5%.
Class B The corresponding figures are 48.5 N/mm^2 and 7.0%.

BRICKS FOR CALCULATED LOAD BEARING BRICKWORK

Bricks for this purpose are classified by strength into classes 1, 2, 3, 4, 5, 7, 10, 15. The average strength ranges from 7.0 N/mm^2 for class 1 to 103.5 N/mm^2 for class 15.

MANUFACTURING DEFECTS IN BRICKS

There are a number of defects that can arise:
under-burning results in a brick of low strength and frost resistance;
over-burning may result in a hard, glassy, distorted brick;
cracking may arise in manufacture. Correct handling and storage will minimise damage to corners, edges and surfaces on facing bricks. Pebbles and other such materials may occur in some bricks.

CALCIUM SILICATE BRICKS

These are manufactured by mixing lime and sand or crushed siliceous gravel. The bricks when formed are subjected to high pressure steam in an autoclave. The lime reacts with the silica to form hydrated calcium silicates which bond the brick. The natural colour of these bricks is near-white but a wide range of colours can be introduced into the raw materials.

In BS 187:1978 the bricks are classified by compressive strength in classes 2, 3, 4, 5, 6, 7. The mean compressive strengths vary from 14.0 N/mm^2 for class 2 to 48.5 N/mm^2 for class 7. The drying shrinkage in all cases is not to exceed 0.04 per cent.

Providing that the correct class of brick is selected they are suitable for all types of brickwork. For example for parapet walls the bricks should be class 3 or higher.
BS 187 gives the recommended minimum class for a wide range of applications. Recommendations for the choice of suitable mortars is also given.

Concrete blocks

Concrete blocks can be made with a wide range of aggregates. These include natural aggregates, foamed slag, air-cooled blast-furnace slag, furnace clinker, crushed clay bricks and tiles, lightweight aggregates and chemically stabilised graded wood particles. The lightweight aggregates include: expanded clay, shale or slate, sintered pulverised fuel ash, pumice and exfoliated vermiculite.

Aerated concrete blocks consist of cement, to which a fine aggregate may be added, which is given an aerated structure by a suitable means and is subjected to high pressure steam in an autoclave. The aerated structure can be achieved by adding fine aluminium powder, which reacts with the alkali produced by the hydration of the cement, to liberate hydrogen gas. Chemical foaming agents can also be used.

Concrete blocks may be described as solid, hollow or cellular. A solid block is one in which the solid material is not less than 75 per cent of the total volume. In a hollow block the holes or cavities pass through the block and the solid material constitutes 50–75% of the total volume. In a cellular block the holes or cavities do not pass through the block but the percentage solid material is the same as that for a hollow block.

TYPES OF CONCRETE BLOCKS

BS 2028, 1364:1968 designates types A, B and C.
Type A blocks have a density of at least 1500 kg/m^3 and have strength designations for minimum average compressive strengths of A(3.5), A(7), A(10.5), A(14), A(21), A(28), A(35); the figures in brackets refer to the compressive strength in N/mm^2. Type A

blocks are suitable for general use in building including use below damp-proof course level.

Type B blocks have a density less than 1500 kg/m^3 and have strength designations of B(2.8) and B(7); the figures in brackets again referring to minimum average compressive strength. These blocks are suitable for general use in building with the proviso that when used in positions below the ground level damp-proof course they should be made of dense aggregate or have an average compressive strength of not less than 7.0 N/mm^2.

Type C blocks are intended mainly for non-loadbearing, internal walls and have a density of less than 1500 kg/m^3. Transverse breaking loads are specified for these blocks rather than compressive strength.

The maximum drying shrinkage of concrete blocks ranges from 0.05% for some blocks of type A to 0.09% for some blocks in type C. For more detail the reader is advised to consult BS 2028.

Gypsum plasters

Gypsum is naturally occurring calcium sulphate dihydrate, $CaSO_4.2H_2O$. In manufacturing gypsum plasters some or all of the water of crystallisation is driven off by heating. The resulting material sets, on the addition of water, by recrystallising as the dihydrate.

CLASSES OF GYPSUM PLASTER

Class A When gypsum is heated to about 150° C about three-quarters of the water of crystallisation is driven off leaving a material which may be written as $CaSO_4.\frac{1}{2}H_2O$ and is thus referred to as a hemihydrate. More correctly the chemical formula would be written as $2CaSO_4.H_2O$. This material is known as plaster of Paris and sets rapidly on the addition of water. It is suitable only for small work such as filling and is the basis of many proprietary fillers.

Class B The rate of setting of plaster of Paris may be retarded by the addition of materials such as keratin to give a retarded hemihydrate plaster, whose full strength is developed in a few hours. BS 1191:Part 1:1973 divides class B plasters into two main types: type 'a' are undercoat plasters for use with sand and type 'b' are final coat plasters to be used neat. Type 'a' plasters are subdivided into browning plaster and metal lathing plaster. The browning plaster is suitable as an undercoat plaster on most clay bricks and concrete blocks. The metal lathing plaster is used as an undercoat on metal laths and woodwool slabs. Type 'b' plasters are subdivided into finish plaster, suitable as a finish over the above undercoats and board finish plaster for one coat work on plasterboard.

Class C These are heated to a sufficiently high temperature to drive off the majority of the water of crystallisation to leave mainly $CaSO_4$ and are thus referred to as anhydrous plasters. The setting time would be very slow but is made more rapid by the addition of accelerators to give a slow steady set. Class C plasters are for finishing coats only and are required to have a greater hardness than Class B finishing coat plasters.

Class D Class D plaster is known as Keene's plaster and is of the anhydrous type. It is purer and harder than Class C plaster and is more easily brought to a smooth, clean finish. It has a gradual set.

LIGHTWEIGHT PREMIXED PLASTERS

These consist of a class B plaster with a suitable lightweight aggregate. The aggregates used are exfoliated vermiculite and expanded perlite. They are used as supplied, only water being added. BS 1191: Part 2:1973 divides the lightweight premixed plasters into type 'a' which are undercoat plasters and type 'b' which are final coat plasters to be used over all the type 'a' undercoats.

The undercoat plasters are subdivided into browning plaster, metal lathing plaster and bonding plaster. The browning plaster is suitable over most bricks and concrete blocks. The metal lathing plaster is used over metal lathing and woodwool slabs. The bonding plaster can be used on a variety of surfaces including concrete, stone masonry, engineering bricks and, with suitable pretreatments, glazed surfaces. The lightweight plasters provide improved adhesion, thermal insulation and fire resistance. They are more resilient than traditional plasters and thus minimise cracking due to movement.

Exercises (*Answers on page 108*)

1. Describe the life cycle of a wood boring beetle.

2. List the conditions that favour attack on timber by
    (i) the furniture beetle;   (ii) the death watch beetle;
    (iii) the house longhorn beetle.
    In each case state the timbers most usually attacked.

3. State the conditions that are likely to cause dry rot.

4. The floor of the dining room of a Georgian farm house consists of oak floor boards on oak joists. Beneath this room is a cellar, the ceiling of which is lath and plaster fixed to the underside of the oak joists. An inspection reveals that the ceiling of the cellar is in a poor state of repair and that the cellar appears slightly damp. Downpipes from gutters are fixed to the outside of one of the dining room walls and disappear below ground in a flower bed. Describe any further investigations to be made and the necessary work to be done to eradicate an insect attack.

5. Referring to question 4 above, part of the floor of the dining room extends beyond the cellars. Removal of a badly disintegrated floor board revealed the floor joists to be buried in debris of bricks, timber and other rubbish. In this part the joists were suffering from fungal attack. Outline the additional investigations and treatment that would be necessary.

In the following questions select the correct options.

6. The exit holes of the house longhorn beetle are:
    (a) circular of 1–2 mm diameter;   (b) circular of 3–4 mm diameter;
    (c) circular of 4–5 mm diameter;   (d) elliptical of 6–10 mm major diameter.

7 Insect damage to timber is due to boring by the:
   (a) beetle;   (b) pupa;   (c) larva.

8 Organic solvent timber preservatives:
   (a) have good penetration;
   (b) are cheap;
   (c) leach out with water;
   (d) are suitable for brushing and spraying;
   (e) do not leach out with water;
   (f) do not cause a fire risk during application;
   (g) may cause a fire risk during application.

9 Softwoods contain:
   (a) tracheids;   (b) vessels;   (c) fibres;   (d) parenchyma cells.

10 The cell characteristic of hardwoods is the:
   (a) tracheid;   (b) vessel;   (c) parenchyma.

11 A type of clay brick has an attractive appearance and a shallow frog. The average compressive strength of these bricks was found to be 50 N/mm^2. The correct classification of these bricks is likely to be:
   (a) a hollow common brick of ordinary quality;
   (b) a solid common brick of ordinary quality;
   (c) a solid facing brick of special quality;
   (d) a cellular facing brick of special quality;

12 The answer to Exercise 11 above requires the following additional information for the classification to be given precisely:
   (a) the density of the brick;
   (b) the volume of the frog;
   (c) the result of a water absorption test;
   (d) the volume of the frog as a percentage of the total volume;
   (e) the results of a soluble salts analysis.

13 Tests on a certain type of clay brick showed that the average compressive strength was 65 N/mm^2 and that the average water absorption was 5.3 per cent. The brick could be classified as:
   (a) engineering class A;
   (b) engineering class B;
   (c) common of ordinary quality;
   (d) internal quality.

14 For constructing an area of decorative paving a suitable type of clay brick would be:
   (a) facing of ordinary quality;
   (b) common of internal quality;
   (c) facing of special quality;
   (d) any brick of suitable appearance.

15 For constructing a parapet wall a suitable type of brick would be:
   (a) a clay brick of special quality;
   (b) a clay brick of ordinary quality;
   (c) a calcium silicate brick of class 2;
   (d) a calcium silicate brick of class 4.

16 A solid concrete block must contain:
   (a) no holes or cavities;
   (b) not more than 25% holes or cavities;
   (c) not more than 30% holes or cavities;
   (d) not more than 50% holes or cavities.

17 For constructing the external leaf of a cavity wall below ground level damp-proof course a suitable type of concrete block would be:
   (a) type A;
   (b) type B(2.8) made with lightweight aggregate;
   (c) type A(14);
   (d) type C.

18. A concrete block for general building work has an average compressive strength of 5.2 N/mm² and a density of 1200 kg/m³. The designations of these blocks is:
    (a) type A(3.5);   (b) type B(2.8);   (c) type A(7);   (d) type B(7).

19. Class A plaster is:
    (a) an undercoat plaster;
    (b) plaster of Paris;
    (c) a rapidly setting plaster;
    (d) a lightweight plaster.

20. Class B plaster is:
    (a) an anhydrous plaster;
    (b) an accelerated hemihydrate plaster;
    (c) a retarded hemihydrate plaster;
    (d) an accelerated anhydrous plaster.

21. Class C plaster:
    (a) is an accelerated hemihydrate plaster;
    (b) is an accelerated anhydrous plaster;
    (c) is an undercoat plaster;
    (d) has a slow, steady set;
    (e) is not as hard as Class A plaster.

22. Lightweight premixed plaster:
    (a) contains a lightweight aggregate;
    (b) should be mixed with sand before use;
    (c) gives a finish harder than Class D plaster;
    (d) is supplied only as undercoat plaster;
    (e) gives good adhesion.

Answers to exercises

CHAPTER 1

1. 173.88 W;
2. 0.89 W/m² °C;
3. −1.48° C;
4. 243.6 W;
5. 10327 W;
6. (a) £10.58, (b) £6.28;
7. (a) 44.66 therms, (b) 1545.3 W, (c) 7.2° C;
8. 11.8° C;
9. 1.37 W/m² °C;
10. 1.47 W/m² °C;
11. 20.4%;
12. 50%;
13. 25%, 23.5%, 22%, 20.4%, 18.8%, 17%;
14. 19.38%;
15. 48.2%;
16. 5.59 W/m² °C, 3.40 W/m² °C, 2.45 W/m² °C, 1.91 W/m² °C, 20.1%, 37.2%, 59.6%, 89.8%.
17. (i) 63%, (ii) 0.008192 kg/kg dry air;
18. 64.6%;
19. (i) 1.4448 kPa, (ii) 79.5%, (iii) 79.2%;
20. (i) 0.00377 kg/kg dry air; (ii) 0.00717 kg/kg dry air; (iii) 49.4%;
21. 1.9368 kPa, 1.9402 kPa;
22. 10.1° C;
23. 48.6%;
24. 66.2%, 13.5° C.

CHAPTER 2

1. 340.3 m/s;
2. 1 ms;
3. 0.041 s;
4. ABCD, 52 μs;
5. 2.72 m;
6. 80 mm;
7. 56.7, 113.3, 170, 226.7, 283.3 Hz;
8. 60 dB;
9. 13.7 dB;
10. 3 dB;
11. 6 dB;
12. 69.8 dB;
13. 76.7 dB;
14. 7;
15. 64.5 dB;
16. 35, 40, 45, 55, 60 dB;
17. 45, 34, 28, 23, 16, 6 dB;
18. 2.5×10^{-8} W/m²;
19. 42.8 dB;
20. 47.1, 49.7, 4.3 dB;
21. 24 kg/m³;
22. 15.2 dB;
23. (c);
24. (d);
25. (a);
26. (a);
27. (c);
28. (a);
29. (b), (d).

CHAPTER 3

1. 20 N/mm²;
2. (b), (c), (d);
3. 32 kN;
4. 45 N/mm²;
5. 400 mm²;
6. (d);
7. 0.00125;
8. 1.2 mm;
9. (a), (c);
10. 212 kN/mm²;
11. 401 N/mm²;
12. 0.26 mm,
13. (b), (d);
14. 140.6 kN;
15. 3.1;
16. see *Fig 1*;
17. see *Fig 2*;
18. (b);
19. (b);
20. 16 kN;
21. (a) see *Fig 3*;
    (b) see *Fig 4*; (c) see *Fig 5*; (d) see *Fig 6*;
22. Reactions: left 1.5 kN, right 2 kN. Maximum shear force = 2 kN.
    Maximum bending moment = 2 kNm. Depth = 116 mm.
    Maximum shear stress = 0.259 N/mm².
23. 2.625 kN.

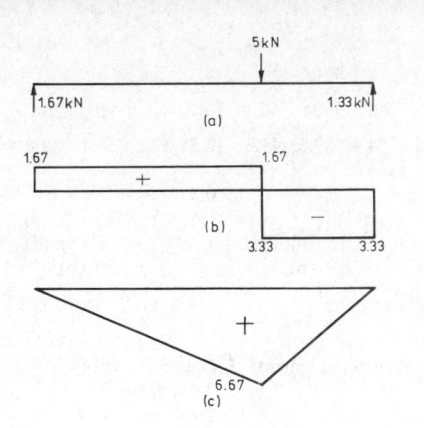

Fig 1 Solution to exercise 16

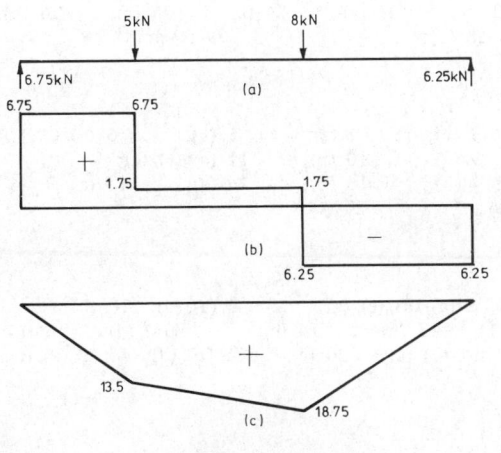

Fig 2 Solution to exercise 17

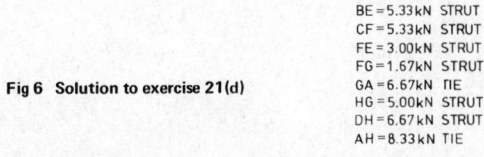

BD = 3 kN STRUT
CD = 5.2 kN STRUT
AD = 2.6 kN TIE

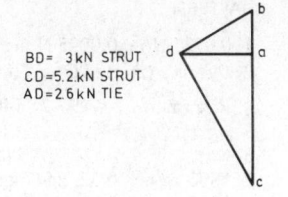

Fig 3 Solution to exercise 21(a)

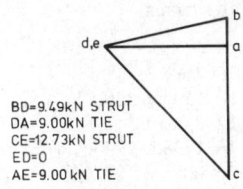

BD = 9.49 kN STRUT
DA = 9.00 kN TIE
CE = 12.73 kN STRUT
ED = 0
AE = 9.00 kN TIE

Fig 4 Solution to exercise 21(b)

BE = 14.00 kN STRUT
EA = 12.12 kN TIE
CF = 6.00 kN STRUT
FE = 8.00 kN STRUT
FG = 4.62 kN TIE
GA = 2.89 kN TIE
DG = 5.77 kN STRUT

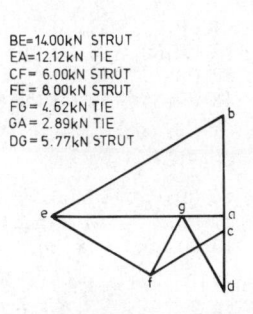

Fig 5 Solution to exercise 21(c)

EA = 6.67 kN TIE
BE = 5.33 kN STRUT
CF = 5.33 kN STRUT
FE = 3.00 kN STRUT
FG = 1.67 kN STRUT
GA = 6.67 kN TIE
HG = 5.00 kN STRUT
DH = 6.67 kN STRUT
AH = 8.33 kN TIE

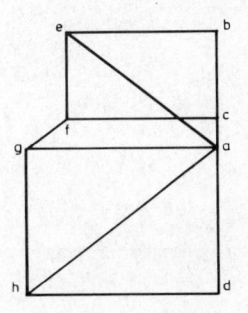

Fig 6 Solution to exercise 21(d)

107

CHAPTER 4

1. 0.0725 N/m, 0.0225 N/m;

2.
Distance (mm)	10	20	30	40	50	60	70	80	90	100
Rise (mm)	297.7	148.8	99.2	74.4	59.5	49.6	42.5	37.2	33.1	29.8

3. No; 5. 0.148 mm; 6. 0.336;
7. 1800 kg/m^3, 0.32, 2647 kg/m^3; 11. (b); 12. (a);
13. (a), (c); 14. (b); 15. (b), (d); 16. (b), (d);
17. (b), (d); 18. (b); 19. (b); 20. (b).

CHAPTER 5

3. C_3S: 4. Low heat Portland cement; 5. Does not comply;
6. Zone 2; 8. Yes; 9. 2.65, 4.2%;
10. 159 kg, 360.5 kg, 25.5 kg; 11. 26%;
12. 0.063 m^3; 14. 55 kg, 500 kg; 15. 0.49, 4.83;
16. 173 kg, 294 kg, 675 kg, 83 kg; 19. (a), (d);
20. (a); 21. (c); 22. (b); 23. (a); 24. (b); 25. (c); 26. (b);
27. (b); 28. (d); 29. (a) or (d).

CHAPTER 6

1. (d); 2. (d); 3. (c); 4. (a); 5. (b); 6. (a), (c), (f);
7. (b), (e); 8. (a), (d); 9. (c); 10. (b); 11. (a), (c), (e);
12. (b); 13. (c); 14. (b); 15. (b); 16. (a); 17. (a);
18. (d); 19. (a), (b), (e); 20. (c).

CHAPTER 7

6. (d); 7. (c); 8. (a), (d), (e), (g); 9. (a), (d); 10. (b);
11. (c); 12. (d) and (e) 13. (b) 14. (c); 15. (a), (d); 16. (b);
17. (a), (c); 18. (b); 19. (b), (c); 20. (c) 21. (b), (d); 22. (a), (e).

Index

Acrylic plastics, 87
Acrylonitrile butadiene styrene, 87
Aggregate-cement ratio, 75
Aggregates for concrete, 69–71
   bulking, 73
   dense, 75
   lightweight, 74
   moisture content, 72–73
Airborne sound insulation, 27–29

Beams,
   bending moment, 43–45
   internal forces, 47–50
   reactions, 39–40
   shear force, 40–43
Bending,
   moment, 43–45
   stress, 47–49
Blocks, concrete, 101–102
Bow's notation, 45–47
Bricks,
   calcium silicate, 101
   clay, 99–101
Bulking of sand, 73

Calcium silicate bricks, 101
Capillarrity, 56–57
Cement, 65–68
Clay bricks, 99–101
Columns, forces in, 50
Compacting factor, 77
Concrete,
   aggregates, 69–73
   blocks, 101–102
   drying shrinkage, 79
   efflorescence, 80
   high density, 75
   lightweight, 74–75
   moisture movement, 79
   strength, 78–79
   sulphate attack, 79
   workability, 76–78
Contact angle, 56
Corrosion, 60–62

Dalton's Law, 9

Damp proof course, 58
Deathwatch beetle, 95
Decibel scale, 22–26
Density,
   bulk, 59
   solid, 59
Dew point temperature, 13
Dry rot, 97
Drying shrinkage, concrete, 79

Electrode potentials, 60
Electrolytic corrosion, 60
Emulsion paints, 90
Epoxy resin, 88

Factor of safety, 38
Frames, 45–47
Fungal attack, 96–98
Furniture beetle, 95

Gypsum plaster, 102–103

Hardwoods, structure, 95
Heat, loss rate, 2
Heating fuels, 3–4
High alumina cement, 68
House longhorn beetle, 98
Humidity, 10–11
   measurement, 14

Impact sound insulation, 29
Insect attack, 95–96
   prevention, 96
   treatment, 96

Lightweight,
   concrete, 74
   plasters, 103
Low heat cements, 67

Masonry cement, 67
Mixing ratio, 10
Modulus of elasticity, 35–37
Moisture,
   content of aggregates, 72–73
   movement, 79

Moisture (*cont.*)
  penetration, 79
Moment of force, 39

Nylon, 87

Oil paints, 88–90
Ordinary Portland cement, 65–66

Paints, 88–90
Percentage saturation, 11
Phenol formaldehyde, 88
Plaster, 102–103
Plastics, 83–88
  properties, 85–86
  structure, 83–85
  thermoplastic, 85
  thermosetting, 85
  types, 86–88
Polycarbonate, 87
Polyester resin, 88
Polyethylene, 86
Polymerisation, 84
Polypropylene, 86
Polystyrene, 87
Polytetrafluoroethylene, 87
Polyvinyl chloride, 86
Polyvinyl fluoride, 87
Polyurethanes, 87

Rapid hardening cement, 66
Relative humidity, 11
  measurement, 14

Saturation vapour pressure, 9–11
Shear,
  force, 40–42
  in beams, 49
  stress, 34
Slump test, 76–77
Softwoods, 94

Sound,
  decibel scales, 22–26
  insulation, 27–29
  velocity, 19–22
  waves, 18
Strain, 35
Stress,
  tensile, 33
  shear, 34
Sulphate attack, 79
Sulphate resisting cement, 67
Supersulphated cement, 68
Surface tension, 54–58

Tensile strength, 37
Thermal transmittance,
  average, 5–8
  definition, 1
  heat loss rate, 5–8
Thermoplastics, 85
Thermosetting plastics, 85
Timber, 93–98
  fungal attack, 96–98
  growth, 93
  insect attack, 95–96
  preservatives, 98
  structure, 94–95
Trusses, 45–47

Ultra high early strength cement, 67
Urea formaldehyde, 88
U-values, 1–8

Vapour pressure, 9–11
V–B test, 77–78

Water-cement ratio, 75
Wet rot, 98
White cement, 67

Young's modulus, 35–37

A checklist of books in the Butterworths Technician Series

BUTTERWORTHS TEC TECHNICIAN SERIES

MATHEMATICS FOR TECHNICIANS 1
FRANK TABBERER, Chichester College of Technology
This is an introduction to mathematics for the student technician, intended especially to cover mathematics at level one in TEC courses (core unit U75/005). The presentation will create an interest in the subject particularly for those students who have previously found maths a stumbling block. There are frequent examples and exercises, with a summary and revision exercise at the end of each chapter.
CONTENTS: Manipulating numbers. Calculations. Algebra. Graphs and mappings. Statistics. Geometry. Trigonometry.
192 pages   May 1978   0 408 00326 X

MATHEMATICS FOR TECHNICIANS 2
FRANK TABBERER, Chichester College of Technology
This covers mathematics at level two in TEC courses (units U75/012 and either U75/038 or U75/039), for those who have completed (or gained exemption from) the work in *Mathematics for Technicians 1*. It includes the alternative schemes of work allowed in the second stage of level two. The clear presentation and systems of examples and exercises, similar to those in the first volume, will enable students to gain a real grasp of the subject.
CONTENTS: Trigonometry (1). Areas and volumes. Statistics (1). Graphs. Trigonometry (2). Equations and graphs. Mensuration. Statistics (2). Introduction to calculus.
156 pages   September 1978   0 408 00371 5

PHYSICAL SCIENCE FOR TECHNICIANS 1
R. McMULLAN, Willesden College of Technology
This is intended for students studying the Physical Science level one unit of programmes leading to TEC certificates and diplomas. The text meets the requirements of the standard TEC syllabus for physical science (unit U75/004), a core unit of courses in building, civil engineering, electrical engineering and mechanical engineering. Attention has been paid to the visual presentation of the text, which is illustrated with diagrams and examples. Important concepts and formulae are clearly highlighted as an aid to learning and revision.
CONTENTS: Introduction. Fundamentals. Force and materials. Structure of matter. Work, energy, power. Heat. Waves. Electricity. Force and motion. Forces at rest. Pressure and fluids. Chemical reactions. Light.
96 pages   May 1978   0 408 00332 4

ELECTRICAL PRINCIPLES FOR TECHNICIANS 2

S. A. KNIGHT, Bedford College of Higher Education

Easy to read and in close conformity with the TEC syllabus, this book is intended primarily to cover TEC unit U75/019, Electrical Principles 2, an essential unit for both telecommunications and electronics students. The text includes examples, worked out for the reader, as well as problems for self-assessment, answers to which will be found at the end of the book. SI units are used exclusively throughout.

CONTENTS: Units and definitions. Series and parallel circuits. Electrical networks. Capacitors and capacitance. Capacitors in circuit. Magnetism and magnetisation. Electromagnetic induction. Alternating voltages and currents. Magnetic circuits. Reactance and impedance. Power and resonance. A.C. to D.C. conversion. Instruments and measurements. Alternating current measurements.

144 pages   May 1978   0 408 00325 1

ELECTRONICS FOR TECHNICIANS 2

S. A. KNIGHT, Bedford College of Higher Education

Provides an introduction to the basic theory and application of semiconductors. It covers the essential syllabus and requirements of TEC unit U76/010, Electronics 2, though some additional notes have been added for clarity. The text includes examples and self-assessment problems.

CONTENTS: Thermionic and semiconductor theory. Semiconductor and thermionic diodes. Applications of semiconductor diodes. The bipolar transistor. The transistor as amplifier. Oscillators. The cathode ray tube. Logic circuits. Electronic gate elements.

112 pages   June 1978   0 408 00324 3

BUILDING TECHNOLOGY 1 & 2

JACK BOWYER, Croydon College of Arts and Technology

These textbooks are primarily intended for the building technician taking TEC B2 construction courses. The clarity of text and illustrations should also, however, appeal to students of architecture and quantity surveying who need a good solid grounding in building construction.

BUILDING TECHNOLOGY 1

CONTENTS: The building industry. Site investigation, setting out and plant. Building elements: practice and materials. The substructure of building. The superstructure of building. Appendix: Building Standards (Scotland) Regulations 1971–75.

96 pages   March 1978   0 408 00298 0

BUILDING TECHNOLOGY 2

CONTENTS: First fixing joinery and windows. Services and drainage. Finishes and finishings. Second fixing joinery and doors. Site works, roads and pavings. Appendix: Building Standards (Scotland) Regulations 1971–75.

96 pages   May 1978   0 408 00299 9

HEATING AND HOT WATER SERVICES FOR TECHNICIANS

KEITH MOSS, City of Bath Technical College

By a system of nearly 200 worked examples, the author describes the routine design procedures for heating and hot water services in commercial and industrial buildings. Primarily intended for student HVAC technicians (TEC sector B3), it will also be useful for other students in sectors B2 and B3, and as a revision aid for experienced HVAC technicians encountering a change from Imperial to SI measurement.

CONTENTS: Heat energy transfer. Heat energy requirements of heated buildings. Heat energy losses from heated buildings. Space heating appliances. Heat energy emission. Heating and hot water service systems. The feed and expansion tank. Three-way control valves and boiler plant diagrams. Steam generation. Steam systems. Preliminary pipe sizing. Circuit balancing. Hydraulic resistance in pipes and fittings. Proportioning pipe emission. Hot and cold water supply. Circulating pumps. Steam and condense pipe sizing. Heat losses using environmental temperature. Medium and high pressure hot water heating.

168 pages   July 1978   0 408 00300 6